Molecular Biology
Biochemistry and Biophysics

5

B. Jirgensons

Optical Activity of Proteins and Other Macromolecules

Second, Revised and Enlarged Edition

With 71 Figures

Springer-Verlag New York · Heidelberg · Berlin 1973

Dr. B. Jirgensons

Professor of Biochemistry

The University of Texas, M. D. Anderson Hospital
and Tumor Institute, Department of Biochemistry
Houston, Texas, USA

The first edition was published in 1969 under the title: Optical Rotatory Dispersion
of Proteins and Other Macromolecules

ISBN 0-387-06340-4 Springer-Verlag New York Heidelberg Berlin
ISBN 3-540-06340-4 Springer-Verlag Berlin Heidelberg New York

ISBN 0-387-04656-9 1st edition Springer-Verlag New York Heidelberg Berlin
ISBN 3-540-04656-9 1. Auflage Springer-Verlag Berlin Heidelberg New York

Preface to the Second Edition

The application of *circular dichroism* (CD) to various problems involving conformation of proteins and other biopolymers is emphasized in this revised and enlarged second edition. The usefulness of CD and ORD in helping to solve structural problems is demonstrated by many examples, and the most essential data are tabulated.

The author is sincerely grateful to the editors of the series Molecular Biology, Biochemistry and Biophysics, especially to Professor GEORG F. SPRINGER, M.D., for their interest in this edition, as well as to the many reviewers for their constructive criticism of the first edition of this book.

Our previously unpublished work reported in this second edition was supported in part by grants from the R. A. Welch Foundation (grant G-051) and U.S. Public Health Service (grant CA-01785).

Houston, September 1973 B. JIRGENSONS

Preface to the First Edition

Great advances have been made in the application of physical methods in the study of the structure of proteins and other biological macromolecules. Optical rotatory dispersion has been successful in solving structural problems, and a vast amount of literature has accumulated on this subject. Several review articles appeared between 1961 and 1965, but significant progress has been made since 1965. Important new studies, especially on the Cotton effects in the far ultraviolet spectrum, have rendered many previous publications obsolete so that a concise monograph should be useful at this time.

The purpose of this writing is to introduce the reader to the use of spectropolarimetric methods and to their applications in Molecular Biology. Proteins are the main objects because of the interest and experience of the writer in this field. A brief survey of nucleoproteins, the glycoproteins, and lipoproteins is also included. It is attempted to interpret the observed phenomena in terms of structure, or more specifically, conformation. Since the complete structure of several crystallized proteins has been elucidated recently by X-ray diffraction and other methods, it is possible to compare the conclusions based on optical rotatory dispersion with those based on X-ray diffraction. This comparison revealed that the rotatory dispersion indeed is able to provide useful information about certain structural features. The limitations of the method are shown, and some of the unsolved problems are exposed.

The presentation is as concise and elementary as possible. The most important theoretical considerations are mentioned, as far as they are useful for interpretation, although experimental methods and results are emphasized. Some recent unpublished results from our laboratory are included. The dispersion constants including recent data on the far ultraviolet Cotton effects, are tabulated.

Because of the enormous quantity of publications in this field, it was impossible to review all reports. Many papers containing important data and interesting ideas, including our own writings, had to be omitted for the sake of better balance. The writer trusts that those authors whose work is not mentioned will be tolerant and forgiving in view of the limitations imposed by the circumstances.

I am grateful to Professor GEORG F. SPRINGER, M.D., for his suggestions and his care in editing this monograph. Our earlier studies on protein structure and the unpublished findings reported in this monograph have been supported in part by a grant CA-01785 from the National Cancer Institute, U.S. Public Health Service, and a grant G-051 from the Robert A. Welch Foundation, Houston, Texas.

Houston, February 1969   B. JIRGENSONS

Contents

Chapter I

The Realm of Proteins. Structural Features.
The Phenomenon of Optical Activity.
Historical Highlights

Proteins, especially those that act as enzymes, are essential ingredients of every living cell. In spite of the spectacular advances in protein structure studies, most of the structures are still either unknown or only partially understood. The situation is even worse as regards the functional aspects, *e.g.* the interplay between the enzymes and other molecules of the living tissue. Nucleic acids are the carriers of genetic information, yet their biosynthesis depends on enzymes. Also, the nucleic acids are found as complex compounds with basic proteins, the histones.

Proteins are defined as macromolecular substances yielding amino acids on hydrolytic decomposition. The fundamental feature of protein structure is the polypeptide chain

$$NH_2 - CH(R_1) - CO \cdot NH - CH(R_2) - CO \cdot NH - CH(R_3) - CO \ldots$$
$$\ldots NH - CH(R_n) - COOH,$$

in which R_1, R_2 .... are the various side chains of amino acids. A complete understanding of structural details is hampered due to (1) the variety of R, (2) the large size of the macromolecule, (3) the intricate arrangement of the polypeptide chain in space, and (4) the fact that many proteins contain carbohydrate, lipid, nucleotide, and other chemically different components.

Great variety exists in the *sizes* and *shapes* of the macromolecules of proteins, and these are important features to be correlated to function. The *fibrous proteins* such as collagen of skin, connective tissue, and bone are used for building certain tissue structures. Some fibrous proteins are also functional, *e.g.* the actomyosin of muscle. Most functional proteins such as enzymes are compact, more or less *globular*. There are globular proteins such as albumin and hemoglobin which function as carriers. In this case, the molecular size is also important. For example, the macromolecules of myoglobin which transfer oxygen into muscle cells, are smaller than the macromolecules of hemoglobin. The internal macromolecular structure of the biologically active proteins is, however, even more important than their size and shape.

The variety of essential enzymes in every living cell is very great, and some of them are present in minute amounts. The isolation and purification of the enzymes, protein hormones, and all other ingredients which are present in low concentrations is a formidable problem. The isolated trace ingredients are often available for study only in quantities of a few milligrams. For this reason, the development of accurate *micromethods* has helped substantially in the chemical investigation of amino acid sequence and physical structural studies. The *spectropolarimetric measurements in the far ultraviolet*, which have been recently developed, and which will be described in this monograph, are important as micromethods in these structural studies.

Several levels of structural complexity can be considered in proteins. The *primary structure* represents the order or sequence of the amino acids in the long polypeptide chain. The term *secondary structure* is used to denote the orderly twisting or bending of the chain, *e.g.* in the form of a cylindrical rod. These extended secondary structures may become more compact in that they are folded back upon themselves, and this folding represents the *tertiary structure*. Finally, two or more polypeptide chains may

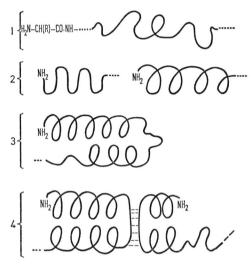

Fig. 1. Schematic representation of primary (1), secondary (2), tertiary (3), and quaternary (4) structures in proteins. In reality, irregular folding of the polypeptide chains is very common, and ordered (*e.g.*, helical) structures are found only in some parts of the macromolecules

combine into larger and more complex units *(quaternary structure)*. While the primary structure is studied chiefly by chemical methods, the physical methods, including optical rotatory dispersion and circular dichroism (see below, and Chapter II), play a decisive role in the disclosure of the higher-order structures. The various structural orders are schematically illustrated in Fig. 1.

The primary structure of a polypeptide chain differs from its secondary and tertiary structures in that all atoms are linked by the strong *primary bonds* (bond energy about 50 to 150 kcal/mole), whereas in the higher order structures several types of *weak secondary bonds* (bond energy 1 to 5 kcal/mole) are involved. Of these, the *hydrogen bonds* between the oxygen of the carbonyl groups $=CO$ and the hydrogen of the imido groups $-NH-$ of the peptide bonds are important. Hydrogen bonds are present in the various types of helices and in the extended β structures. The *nonpolar hydrophobic bonds* between the hydrocarbon residues of the amino acid side groups, such as $CH_3-CH_2-CH_2\ldots$, C_6H_5- now are known to be of considerable importance. Although they are weak, their large number along the polypeptide chain produces significant effects. The hydrophobic bonding is caused by the same forces which hold together soap molecules in relatively large packages, the so called micelles. The hydrophobic "tails" of the soap molecules tend to coalesce, because they do not have affinity to the polar water; and in these considerations, the structure of water is as important as the composition of the protein (KAUZMANN, 1959; TANFORD, 1961; MARTIN, 1964).

This leads to the question: what causes the polypeptide chain to assume certain more compact forms? In the process of biosynthesis, the polypeptide chain is formed by linear linking of the activated amino acids on the surface of a ribosome (see reviews of FRUTON, 1963, and GROS, 1965). After the chain is completed, it is released, assuming then a certain secondary and tertiary structure. What causes the chain to twist and fold? The concept now prevails that the *amino acid composition and sequence or primary structure determines the higher-order structures.* The *ratio of the polar to nonpolar side groups* of the constituent amino acids is very important. When polar side chains predominate, the macromolecule stretches out. For example, the histones, which are composed chiefly of the basic amino acids with polar side chains (lysine, arginine), appear as highly extended chains in water, whereas polypeptide chains containing an excess of nonpolar residues fold into compact globular entities. In the folded macromolecule, most of the nonpolar side chains are *inside* and most of the polar groups are outside of the unit, as it has been ascertained by X-ray diffraction (see *e.g.* PHILLIPS, 1963). Not only the protein is important in these considerations but also the *environment* in which it is placed. The histones are highly extended in pure water, but they are folded up at sufficiently high concentrations of electrolytes, *e.g.* at the DNA chains containing ionized phosphate (JIRGENSONS and HNILICA, 1965). Of course, these considerations do not explain all the phenomena observed. They do not tell us why of the many possible secondary structures some helical forms are more common than others or why sometimes the folded chain is unordered. Extensive studies of several groups of investigators have shown that the spatial arrangement in the large molecules depends on the bond length, bond angles, and strain energy involved in joining the amino acids into various polypeptide structures. Some of the structures are more probable and stable than others, because in certain arrangements there is a minimum of strain. The *α-helix* of PAULING and COREY in which there are 3.6 to 3.7 amino acid residues per turn of the helix, is such a stable structure (PAULING and COREY, 1951, 1954; PAULING, COREY, and BRANSON, 1951). The α-helix has been discovered in many fibrous and globular proteins, chiefly by X-ray diffraction and optical activity studies.

The disorganization of the secondary and tertiary structures is called *denaturation* (KAUZMANN, 1959; JOLY, 1965). If a protein is denatured and the denaturing agent (*e.g.* urea or acid) removed, the protein often returns into its original native state. This happens even after destruction of disulfide bridges, *e.g.* by reduction; reoxidation by the oxygen of air results in *"renaturation"*, *i.e.* return to the original secondary, tertiary, and even quaternary structure. From these observations it has been concluded as already stated, that the primary structure, which is determined in biosynthesis by the genetic code, determines the higher-order structures (ANFINSEN, 1962; WHITE, 1961; WHITNEY and TANFORD, 1965). Optical activity measurements have been very helpful in these discoveries, as will be shown in later chapters.

The optical activity or chiroptical methods are based on the interaction of mono-chromatic polarized light with the valence electron shells of the atoms of a substance. The result of this interaction is the phenomenon of rotatory dispersion; it depends on the *arrangement of the atoms in space* and on the wavelength of the light which strikes them. Thus a polypeptide chain in helical arrangement yields effects which differ strongly from those displayed by a disordered chain.

The optical rotatory dispersion phenomenon was discovered more than 150 years ago by the French scientist BIOT. He observed the behavior of light of different color passing through quartz crystals, and he noticed that the angles of rotation of the plane of polarization were inversely proportional to the squares of the wavelengths. BIOT was also the first who observed the rotatory dispersion of organic compounds. "The evidence which I have just reported seems to me sufficient to prove that the law of rotation of different simple rays is the same for the liquid sugar as for rock crystal and oil of turpentine... From this one can infer with great probability that the law is a general one..." A clearer understanding of the phenomena was provided later by A. FRESNEL (1824). "Polarized light is that in which the transverse oscillations take place constantly in one direction... The act of polarization does not consist in creating transverse vibrations, but in decomposing them along two fixed directions at right angles to one another..." FRESNEL (1824) again clearly envisioned also the causes for the rotation of plane of polarization. "There are certain refracting media, such as quartz in the direction of its axis, turpentine, essence of lemon, etc., which have the property of not transmitting with the same velocity circular vibrations from right to left and those from left to right. This may result from a peculiar constitution of the refracting medium or of its molecules, which produces a difference between the directions right to left and left to right; such, for instance, would be a helicoidal arrangement of the molecules of the medium, which would present inverse properties as these helices were dextrogyrate or levogyrate." Two important aspects of optical rotation are pointed out in this quotation: (1) the importance of the *refractive index*, and (2) the *asymmetry of the refracting medium*. In a quartz crystal the asymmetry is due to the structure of the whole crystal (the SiO_2 molecules themselves are not asymmetric), whereas in a liquid system *the molecules* of the liquid, or of some of its components, must be asymmetric in order to affect the beam of polarized light.

Significant contributions to the study of stereochemistry of organic compounds were made in the middle of the last century (1848 to 1860) by L. PASTEUR. "When we study material things of whatever nature, as regards their forms and the repetition of their identical parts, we soon recognise that they fall into two large classes of which the following are the characters. Those of the one class, placed before a mirror, give images which are superposable on the originals; the images of the others are not superposable on their originals, although they faithfully reproduce all the details. A straight stair, a branch with leaves in double row, a cube, the human body—these are of the former class. A winding stair, a branch with the leaves arranged spirally, a screw, a hand, an irregular tetrahedron—these are so many forms of the other set. The latter have no plane of symmetry." Two important spatial arrangements were envisioned by PASTEUR: the *irregular tetrahedron* and the *screw* or *helix*. In the last part of the 19th century, attention of chemists was directed chiefly to the irregular tetrahedron, *i.e.* the asymmetric carbon, as the principles of stereochemistry were established by LE BEL and VAN'T HOFF [1] (1874). This is related to an unprecedented upsurge of classical organic chemistry. At that time, the organic chemist was concerned with the structure of small molecules, such as hydroxy acids, amino acids, sugars. Optical isomerism was mostly tested by determining the rotatory power at a single wavelength, usually with yellow sodium light (wavelength 589 mμ). The rotatory dispersion was neglected, and it is believed that the use of the Bunsen burner, which was very convenient in providing the yellow sodium flame, was one of the reasons why the vast majority of optical activity data were obtained at this wavelength. This tradition continued far into the twentieth century.

The study of the structure of proteins and other macromolecules was started successfully only at the beginning of the twentieth century. Of crucial importance in the development of sound guidelines in these endeavors were several factors: (1) the idea of STAUDINGER that proteins and other natural polymeres (starch, cellulose, rubber) are *macromolecular substances* (he published a monograph on the early work in 1932); (2) application of the X-ray diffraction method to the study of biopolymers (*e.g.* ASTBURY, 1933); and (3) the results of analytical ultracentrifugation of protein solutions (SVEDBERG and his colleagues, 1925 *et seq.*; early work was outlined in SVEDBERG and PEDERSEN, 1940). The first point is now a generally accepted truth, but it was not so even in the late thirties. From 1920 to about 1935, the majority of chemists believed that the particles of proteins and other organic colloids are not large molecules but micelles, *i.e.* that the entities in a solution are composed of many small molecules linked by secondary bonds. In 1922, STAUDINGER proclaimed the idea that biocolloids, like starch, or proteins, or rubber, are macromolecular substances, *i.e.* that their molecular weight is very high (over 10,000), and that the entities in solution are not micellar aggregates but *large molecules* in which all atoms are linked by strong primary bonds. This is an example of the clear intuitive vision of a genius, since the experimental basis for such a statement at that time was very meager. Supporting data for the new concept were provided by STAUDINGER and his school (*e.g.* STAUDINGER, 1932, 1950, 1961). In the protein field, the ultracentrifugation results were important (SVEDBERG and PEDERSEN, 1940). They showed that, with respect to molecular size, the solutions of many proteins are homogeneous and that the molecular weights are high. In most of these homogeneous proteins, such as albumin and hemoglobin, the entities in the solutions were stable; for example, the molecular weight did not change upon addition of salts, dilution, or moderate variation of temperature. Thus the entities probably were macromolecules, because in the case of micelles such as aqueous solutions of soap the particles disaggregate and aggregate readily, depending on the solvent and temperature.

The first ideas about the higher-order structures in proteins emerged from the X-ray diffraction studies of solid fibrous proteins, chiefly by ASTBURY (1933). A certain order in the samples was apparent and, interpreted in the perspective of the macromolecular concept, yielded the first insight into the higher-order structures. Later studies of PAULING and COREY (*e.g.* 1951, 1953, 1954) were very important.

---

[1] It should be mentioned that the ideas about the asymmetric positioning of atoms in space were not accepted by all chemists. Thus the noted German organic chemist KOLBE qualified the theory of VAN'T HOFF as "fancy trifles ... totally devoid of any factual reality and ... completely incomprehensible to any clear minded researcher" (see VAN'T HOFF, 1874, transl. by G. F. SPRINGER, 1967).

These investigators compared the X-ray diffraction data with models calculated and constructed on the basis of these data, including the bond lengths and bond angles of amino acids and small peptides, and they were able to formulate several secondary structures, such as the *α-helix* and the *parallel* and *antiparallel β forms*. These are illustrated in Figs. 2 and 3. The polypeptide chain, when fully extended, is a zigzag structure because of the angles formed between the $C-N-C-C-N$-atoms in the chain. The $C-N$ bond has a partial double-bond character and the four atoms of the peptide link and the two adjacent α-carbon atoms all are in the same plane. A long

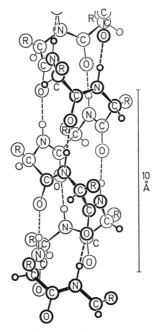

Fig. 2. Schematic drawing of the α-helix after PAULING and COREY (1953)

chain, remaining in the same plane, can fold back on itself thus forming sheet-like structures which can be stabilized by *intra*chain hydrogen bonds between the CO and NH groups in the loops. Sheet-like β structures can be formed also of many chains running parallel or antiparallel in the same plane and forming *inter*chain hydrogen bonds between the CO and NH groups. The actual structures depend on the primary structure which should provide appropriate conditions for the accommodation of the side chains. The antiparallel pleated sheet β structure, as found *e.g.* in silk and some globular proteins, has the CO . . . HN bond perpendicular to the chain axis and the axial repeat is 6.5 Å. In the helix, two asymmetry features are obvious: the *asymmetric carbon atoms,* which can be thought of as being in the center of an imaginary tetrahedron and surrounded by four different groups of atoms, and the *asymmetry of the helix itself.* Two stereoisomeric α-helices can be formed from any chain by twisting it clockwise and counterclockwise. These *right-* and *left-handed* helical

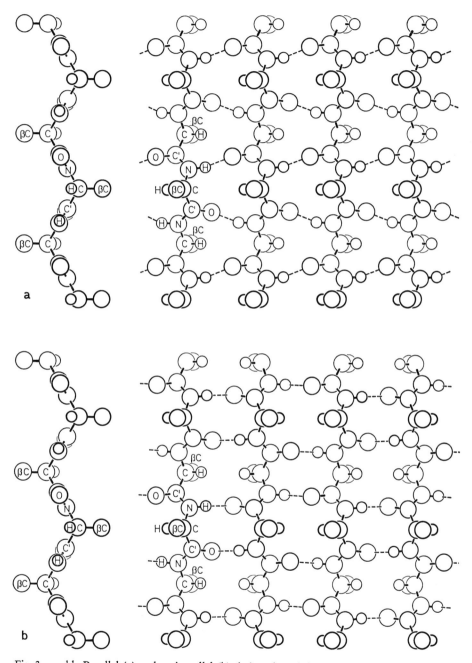

Fig. 3 a and b. Parallel (a) and antiparallel (b) chain "pleated sheet" or β structures of extended polypeptide chains, after PAULING and COREY (1953)

structures are not superimposable and their optical characteristics are opposite. The bond angles between the atoms of the chain backbone provide the condition for this structure which has a minimum of strain energy. The helical winding makes the chain shorter than it is in the fully extended form. The pitch of the α-helix is about 5.4 Å, which amounts to 1.48 Å per residue, and the side chains extend away from the helix axis. Each NH group of the backbone is linked by a hydrogen bond to the carbonyl group of the third distant peptide bond in the helix [2]. These hydrogen bonds have the length of approximately 2.8 Å, and they are nearly linear. According to PAULING (1960), a deflection of the NH . . . O bond by 6° from a straight line at the hydrogen atom produces 0.1 kcal/mole of strain energy. Although the right-handed α-helix has a minimum of strain energy, the helical structures found in proteins are often distorted (e.g. PHILLIPS, 1968). Moreover, the stability of the helices depends not only on hydrogen bonding but also on the interactions between the closely situated side chains. Since the hydrogen bonds of the helix may be affected by competitive interactions with the water in which the protein is dissolved, the helical structures are usually found inside the large protein molecules or in protein fibers of low moisture content. However, spiral structures differing strongly from the α-helix also have been found in proteins; thus the helix found in collagen differs strongly from the α-helix. The stability of the ordered structures has been reviewed by SCHELLMAN and SCHELLMAN (1964) and FASMAN (1967). A clear, accurate, beautifully illustrated account of the various structures has been provided by DICKERSON and GEIS (1969).

The study of proteins and other biopolymeres by the method of optical rotatory dispersion was rather neglected until about 1955. The same is true of the rotatory dispersion of smaller molecules, as pointed out in the monographs of DJERASSI (1960) and CRABBÉ (1965). The reasons for this were: (1) the intricacy of the interactions between light and the electrons that hampered the development of a theory which would permit a straightforward interpretation of experimental data, (2) the dominating influence of the X-ray diffraction method, and (3) lack of suitable instrumentation. The revival of the optical rotatory dispersion method occurred between 1955 and 1960, when new experimental and theoretical tools became available, and when *simple model substances (polyamino acids)* were synthesized (see FASMAN ed., 1967). The models helped the protein chemist considerably in the interpretation of the data obtained with proteins, because the secondary structure of these simpler polymers could be established by several independent physical methods. Progress was accelerated by further advances in experimental techniques that opened up the far ultraviolet spectral zone for study. A spur to the application of the method to proteins was provided by the agreement between X-ray diffraction results and conclusions based on optical rotatory dispersion (see Chapter VII).

The applications of the optical rotatory dispersion method in the studies of protein structure are described in many review articles, e.g. of TODD (1960), URNES and DOTY (1961), JIRGENSONS [1961 (1)], SCHELLMAN and SCHELLMAN (1961), RIDGEWAY (1963), FASMAN (1963), and SCHELLMAN and SCHELLMAN (1964). The important recent advances in far-ultraviolet spectropolarimetry, to about 1965/66, have

---

[2] This cannot occur with proline and hydroxyproline, because of the absence of hydrogen at the peptide nitrogen. For this and other reasons the α-helix content in proteins of high proline content is very low.

been reviewed briefly by HARRINGTON, JOSEPHS, and SEGAL (1966) and by YANG [1967 (1, 2)].

Several books have appeared on the optical rotatory dispersion method and its applications in organic chemistry (DJERASSI, ed., 1960; CRABBÉ, 1965; SNATZKE, ed., 1967). The classical outline of LOWRY (1935) has been reprinted (1964). However, in these books, proteins have received only limited attention.

While in 1955—1965 optical rotatory dispersion (ORD) was the major chiroptical method in structural studies, the *circular dichroism* (CD) method emerged in the late sixties as a new and powerful tool (see next chapter). Circular dichroism was studied earlier chiefly by COTTON (1896) and this method was revived for the same reasons as apply to ORD. More recent applications of CD in organic chemistry have been reviewed by VELLUZ, LEGRAND, and GROSJEAN (1965), and the first applications to structural studies of biopolymers were treated by HOLZWARTH and DOTY (1965), BEYCHOK (1966, 1967) and TIMASHEFF et al. (1967). Among recent reviews are those of URRY (1970) and TINOCO and CANTOR (1970).

Chapter II

The Phenomena of Optical Activity.
Terms and Definitions. Theoretical Considerations.
The Drude and Moffitt Equations

1. Refraction of Monochromatic Polarized Light

When a ray of monochromatic polarized light strikes a solution, several events may occur: (1) reflection on the surface, (2) refraction, (3) rotation of the plane of polarization, and (4) absorption. Reflection can be minimized by proper optical arrangement and will not be discussed here. However, the other three events are important in the problems to be considered. *Refraction* is caused by slowing of the ray. In the language of the electromagnetic theory of light and electronic structure of matter, refraction is caused by interaction of the propagated electromagnetic field with the oscillating electrons of matter. A propagated electromagnetic field, according to the classical theory of MAXWELL, is characterized by an *electric vector E* and a *perpendicular magnetic vector H*. If the light is monochromatic, *i.e.* has a definite wavelength λ or frequency ν ($\lambda = 1/\nu$), the electric vector E at any point is expressed as $E = E_0 \cos w\,t$, where E_0 is the maximum amplitude of the wave, $w = 2\pi\nu$, and t is time. A similar expression, $H = H_0 \cos w\,t$, is valid for the magnetic vector H, which shows that E and H, though perpendicular to one another, are in phase. In a refracting medium the wave is slowed and the amplitude of the electric field component E is diminished, and, according to the theory, the decrease is proportional to the factor $3/(n^2 + 2)$, where n is the refractive index. The decrease in the magnitude of E is caused by induction of dipole moments in the refracting molecules, and the disturbance depends on the number and distribution of the electrons. In a nonabsorbing isotropic medium, *e.g.* a solution containing only symmetric molecules, refraction will be the only major event. All secondary waves emerging from the solution will be polarized in the same direction as the incident wave, *i.e.* no optical activity will be observed. In order to generate optical activity, the substance interacting with the light should be able to affect not only the electric but also the magnetic vector (see *e.g.* SCHELLMAN, 1958).

2. Optical Rotation

In the case of a linearly (or plane) polarized ray of a definite wavelength λ, the electric field associated with it will oscillate sinusoidally along a single direction in space. As already pointed out by FRESNEL (see Chapter I), the variable field vector, or component of the ray, can be considered as a resultant of two equal vectors corresponding to a right circularly polarized wave E_r and to a left circularly polarized wave E_l. In an optically inactive medium, the vectors E_r and E_l can be imagined as forming equal angles with the direction of vibration (Fig. 4). In a medium containing *asymmetric molecules*, however, E_r and E_l are traveling at different speeds, the angles between them and the directional plane being different. This causes a phase shift. The resultant ray still will be plane polarized, but its plane of polarization will be rotated by an angle α. This rotation occurs because the molecular asymmetry results in different polarizability of electron shells in different directions. If the events are studied at a wavelength *far from the absorption bands*, the difference can be explained as a

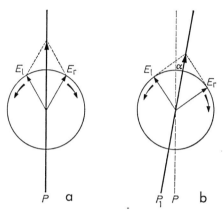

Fig. 4 a and b. Circularly polarized and plane polarized light in optically inactive (a) and optically active (b) medium. The arrows show the direction of the electric vector. In circularly polarized light the electric vector and the plane of polarization rotate either clockwise or counterclockwise. In the optically inactive medium (case a) the left circularly polarized wave E_l travels with the same velocity as the right circularly polarized wave E_r. The resulting vector is the sum of $E_l + E_r$, yielding plane polarized light in the plane P. In the optically active medium (case b) the vectors E_l and E_r travel with different velocities, resulting in a rotation of the plane of polarization by an angle α.

The velocity with which the vectors travel correspond to the *refractive indices*. The velocity of light in vacuum, c, is always greater than its velocity in a substance. If the vector E_l travels with velocity c_l and the vector E_r with velocity c_r, the ratios c/c_l and c/c_r are the refractive indices. In an optically active medium, c/c_l differs from c/c_r and we have *circular birefringence*

difference in the refractive indices for the directions "right" and "left", i.e. n_r and n_l. The angle of rotation of the plane of polarization, α, is proportional to the difference $n_l - n_r$, and it is obvious that it can be either *positive* or *negative*. Moreover, at a constant wavelength, λ, the angle of rotation must be proportional to the *number of the asymmetric molecules*. In the instance of a solution, this factor can be varied in two ways: either by *variation of the optical path* through the solution or by *variation of concentration* of the optically active solute.

3. Specific Rotation

Taking into account the optical path and concentration, the optical rotation at a constant wavelength, λ, is expressed as *specific rotation* $[\alpha]_\lambda$, *viz.*:

$$[\alpha]_\lambda = 100\ \alpha/c\ l \qquad (1)$$

in which α is the rotation in degrees, c is the concentration of the optically active solute in grams per 100 ml, and l the optical path through the solution in decimeters.

The term *molar rotation*, $[M]_\lambda$, takes into account also the molecular weight of the solute, and is defined as

$$[M]_\lambda = [\alpha]_\lambda M/100, \tag{2}$$

where M is the molecular weight.

4. Corrected Mean Residual Specific Rotation

In the case of proteins, it appears useful to consider the *mean residual weight* of the constituent amino acids, $\bar{M}$, instead of the molecular weight of the protein. $\bar{M}$ is obtained by dividing the molecular weight of the protein by the number of amino acid residues. In addition, to include the refractive properties of the solvent, it is now general practice to consider the factor $3/(n_\lambda^2 + 2)$, where n_λ is the refractive index of the solvent. Thus the corrected mean residual specific rotation, denoted by either $[m']_\lambda$ or $[R']_\lambda$, is expressed as

$$[m']_\lambda = [R']_\lambda = [\alpha]_\lambda\, 3\, \bar{M}/100\, (n_\lambda^2 + 2)\,. \tag{3}$$

The $\bar{M}$ values of most proteins are within the range of 100 to 120, hence the $\bar{M}/100$ values vary between 1 and 1.2. The $3/(n_\lambda^2 + 2)$ values are fractions not differing much from unity. The whole correction factor $3\,\bar{M}/100\,(n_\lambda^2 + 2)$ at the longer wavelengths in the visible spectrum amounts to approximately 0.9. It is, however, considerably smaller in the far ultraviolet (see Sect. 10). A more detailed specification may include the temperature of measurement, and the appropriate symbols then are $[\alpha]_\lambda^t$ or $[m']_\lambda^t$. The optical rotation of protein solutions is usually determined at ambient room temperatures of 20 to 30° C. The results are not significantly affected by small changes in temperature. However, the constancy of temperature may be important, *e.g.* when the kinetics of denaturation or the kinetics of some enzymic reaction are studied. The subscript λ at the symbols of n, $[m']$, or $[\alpha]$ is often omitted, *i.e.* the specific rotation is often denoted by $[\alpha]$.

5. Optical Activity and Absorption of Light

The statement that optical activity is caused by a difference of refractive indices of an asymmetric medium is true only for solutions in which the absorption of light is negligible. If the events are considered *near an absorption band*, the *difference in extinction coefficients* also must be taken into account. When the absorption becomes significant, the medium exhibits *circular dichroism* (see below), and the light is elliptically polarized. In or very near an absorption band, the interactions between the polarized ray and the electrons of the matter are maximal. Not all of the atomic groups (or electronic configurations), however, interact with the light with the same intensity. The atomic groups which are distinguished in these interactions are called *chromophores*, and it is noteworthy that the *electronic environment of these groups within the molecule is as important as the structure of the groups themselves.* In other words, some chromophores may even be symmetric but the interactions may occur in an *asymmetric environment* in the molecule (KAUZMANN, WALTER, and EYRING, 1940; MOSCOWITZ, 1960; BEYCHOK, 1966).

As indicated in Sect. 1, in an optically active medium both electric and magnetic vectors are involved. According to the laws of physics, which have been tested with such macroscopic devices as electric generators and other machines, the electric and magnetic phenomena are closely interrelated; and this applies also to the submolecular events. In an optically active substance, this interrelationship depends on specific structural features of the medium, and it can be expressed as follows:

$$\mu = \alpha' E' - \frac{\beta}{c} \cdot \frac{\partial H}{\partial t}, \text{ and } m = \frac{\beta}{c} \cdot \frac{\partial E'}{\partial t},$$

in which μ is the electric moment, α' the molecular polarizability, E' the electric vector, c the velocity of light, H the magnetic vector, t time, and m the induced magnetic moment; β is the structural parameter which links the electric and magnetic effects and is responsible for the optical rotation (see SCHELLMAN, 1958). In the same connection, it is noteworthy that optical activity can be induced by an external magnetic field (Faraday effect, see Sect. 12).

The important absorption bands of proteins are in the *far* ultraviolet part of the spectrum, and they became accessible to direct measurement but recently. The optical activity measured at certain wavelengths in the visible or near ultraviolet part is relatively weak. Moreover, it is a *composite effect* caused by electronic transitions of many chromophores.

Besides the processes of refraction, optical rotation, and absorption, irradiation of a solution by light of high intensity and frequency may lead to permanent chemical change of the specimen. Such photochemical reactions do not occur in the ordinary optical measurements, because the intensity of the most active ultraviolet light is not high enough in the ordinary equipment used for optical rotation measurement. Other phenomena, such as light scattering and fluorescence, also have been omitted from these considerations.

6. Optical Rotatory Dispersion. The Drude Equation

In the previous sections, it was assumed that the wavelength of the monochromatic polarized light was not changed. A variation of the wavelength results in a drastic change of the optical activity of a given substance, and *the change is greater the closer one approaches the absorption band*. The measurements are started at some point in the long wave zone and continued toward the short wave end, now usually with the aid of a recording spectropolarimeter. The values of the optical activity, α, are plotted directly against wavelength, λ, whereby curves like those shown in Fig. 5 are obtained. Several essential features of these protein curves are noteworthy. First, the curves are in the negative part of the chart, i.e. the *proteins are levorotatory*. Second, the *levorotation increases with decreasing wavelength according to a power function*. In the examples shown, curve *1* represents a case in which the protein is more asymmetric than in case *2*, as expressed in a more drastic change with decreasing wavelength. At a certain long wavelength, the rotatory power of protein 1 is less than that of protein 2, whereas at the short wavelengths the opposite is true. It was realized early that a rotatory dispersion curve express the asymmetric structures more completely than optical activity measurement at a single wavelength, for example at the sodium line of $\lambda = 589$ nm.

The electronic theories of optical rotatory dispersion have been developed and reviewed by DRUDE (1900), ROSENFELD (1928), LOWRY (1935), KIRKWOOD (1937), KAUZMANN, WALTER, and EYRING (1940), MOFFITT (1956), SCHELLMAN (1958), and others. Although different models of interaction have been used by different authors, the general relationships were found to be the same or very similar.

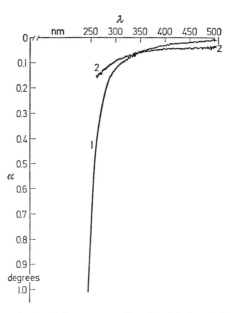

Fig. 5. Dependence of optical activity on wavelength of light of fumarase (curve *1*) and elastase (curve *2*). On the abscissa is plotted the wavelength, λ, and on the ordinate the optical activity. The concentration was 0.04%, and tube length was 5.0 cm

According to DRUDE (1900); the optical activity at a wavelength λ depends on a sum of electronic transitions, i:

$$[\alpha]_\lambda = \sum_i K_i/(\lambda^2 - \lambda_i^2) \tag{4}$$

where λ_i is the wavelength of any of the transitions resulting in optical activity, and K_i are constants proportional to the rotational strength (number of vibrators, and other factors) to be discussed later.

Experimental studies with many different substances have shown that the data can be accommodated by a simplified form of Eq. (4), *i.e.*

$$[\alpha]_\lambda = A/(\lambda^2 - \lambda_c^2) \tag{5}$$

in which λ_c is the wavelength corresponding to *one* dominating interaction ($\lambda_c = \lambda_i$), and A is another constant characteristic for each system. This simplified Drude equation has been found valid also for solutions of colorless proteins when the rotatory dispersion was measured in the visible and near ultraviolet spectral zones (*e.g.* between 350 and 600 nm), that is, *far from the absorption bands*. Among the first measurements on proteins which were analyzed this way were those published by HEWITT in 1927. The Drude plots are now performed almost exclusively according to YANG and DOTY (1957), *i.e.* by plotting $[\alpha]_\lambda \cdot \lambda^2$ against $[\alpha]_\lambda$, since Eq. (5) can be rearranged easily into

$$[\alpha]_\lambda \lambda^2 = [\alpha]_\lambda \lambda_c^2 + A . \qquad (6)$$

The λ_c, the so called *dispersion constant,* is then obtained from the slope of the straight line, and A, the *rotatory constant,* is provided by the intercept on the ordinate. Examples of this type of plotting are shown in Fig. 6 [1].

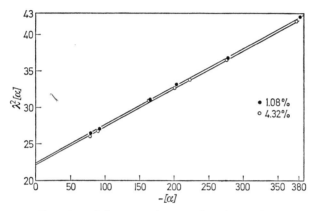

Fig. 6. The rotatory dispersion of chymotrypsinogen, plotted according to YANG and DOTY (1957). The dispersion constant, λ_c, is found from the slopes of the straight lines. [From JIRGENSONS, 1958 (1)]. It is expressed in microns

While the interpretation of the significance of the constant A appeared to be complicated, it was realized early that the dispersion constant, λ_c, is able to provide some limited amount of information on the higher-order structures in proteins. From 1957 to 1960, the λ_c values of many proteins were determined, mostly by the writer, and a complete compilation of these data can be found in the review article of URNES and DOTY (1961). The λ_c values of proteins were found to be in the range of 180 to 290 nm, the higher values indicating a higher degree of structural asymmetry in the macromolecules. However, this appeared to be a rather crude approximation, and the need for further refinement of the method was obvious. A certain magnitude of λ_c, *e.g.* $\lambda_c = 280$, does not mean that the rotatory dispersion is determined by a single

---

[1] In these Drude plots the wavelength is expressed in microns, and the λ^2 values needed for such work are now tabulated for many wavelengths (FASMAN, 1963). In the case of chymotrypsinogen shown in Fig. 6, the intercept of the upper line is 22.2 and the slope is 0.054. The dispersion constant, λ_c, is found by taking the square root of 0.054, *i.e.* $\lambda_c = 0.232\ \mu$. Usually, however, the λ_c values are expressed in nm, *i.e.* the λ_c of chymotrypsinogen is 232 nm (± 1).

transition at that wavelength. The aromatic absorption band of proteins at 275 to 282 nm certainly is of minor importance in the rotatory dispersion observed in the visible and near ultraviolet spectral zones.

7. The Moffitt Equation for the Rotatory Dispersion of Helical Structures

The theoretical endeavors of MOFFITT (1956) in correlating rotatory dispersion of proteins to the helical secondary structure have been very stimulating. It was assumed that the *periodic disposition of the constituent atomic groups provides a system of chromophores in which the electronic transitions occur parallel and perpendicular to the screw axis of the helix.* It was deduced that, for wavelengths far from the absorption bands, the rotatory power dependence should be expressed by

$$[m']_\lambda = \sum_i a_i \lambda_i^2/(\lambda^2 - \lambda_i^2) + \sum_i b_i \lambda_i^4/(\lambda^2 - \lambda_i^2)^2 \,, \tag{7}$$

where λ_i is the dispersion constant, as in the Drude equation (4), and a_i and b_i are two other constants. Of these, the b_i is specifically related to the helix content.

The Moffitt theory has been justly criticized as being on oversimplification. Also, it has been pointed out by KAUZMANN (1957) and by MURAKAMI (1957) that coupling effects of identical chromophores other than the helix can yield similar theoretical relationships to those expressed in Eq. (7). In spite of these objections, the Moffitt theory played a significant part in the application of optical rotatory dispersion to the study of the higher-order structures in synthetic polyamino acids and proteins.

Since Eq. (7) is not directly applicable to the treatment of experimental data, MOFFITT and YANG (1956) modified it to the simplified form

$$[m']_\lambda = a_0 \lambda_0^2/(\lambda^2 - \lambda_0^2) + b_0 \lambda_0^4/(\lambda^2 - \lambda_0^2)^2 \,. \tag{8}$$

This still has *three adjustable parameters*, a_0, b_0, and λ_0, and its applicability was tested by plotting $[m']_\lambda (\lambda^2 - \lambda_0^2)$ against $1/(\lambda^2 - \lambda_0^2)$, and trying to *find the λ_0 which would yield the best straight line.* Synthetic polyamino acids which were known to be helical were used as examples, and their optical activity was determined in the visible and near-ultraviolet spectral zones. One such example is shown in Fig. 7. According to these data, the value of 2120 Å or 212 nm for λ_0 gave the best straight line, and thus was suggested as a constant in all determinations. The variable parameter a_0 was found from the intercept and the value of b_0 *from the slope* of the straight line. While the magnitude of a_0 was found to be relatively sensitive to the variation of solvent and side chain composition of the helical polymer, the b_0 values of the polymers, which were known to be α-helical from studies with other methods, were near −630. For example, poly-γ-benzyl-L-glutamate in dioxane yielded an a_0 of 135 and a b_0 of −630; the same polymer in a chloroform solution had an a_0 of 250 and a b_0 of −625. In the case of poly-α-L-glutamic acid the b_0 in aqueous solutions of pH 4.5 was −630, whereas in neutral or weakly alkaline aqueous solutions it was near zero. Again, it was ascertained by other independent methods that in the acid solutions this polymer is helical and that in neutral or alkaline solutions it is disorganized. On the basis of these findings, it was suggested that the α-*helix content* in polymers and proteins with a partial helical order could be obtained by linear interpolation, *i.e.* that the helix content, in per cent, is 100 $b_0/630$. During the last 10 years the helix content in many proteins has been determined this way in many laboratories. However, in these estimates it was assumed that the α-helix and flexible disorder are the only structures. This was later found to be an oversimplification [see *e.g.* JIRGENSONS, 1958 (1), 1961 (1); TANFORD, DE, and TAGGART, 1960; SCHELLMAN and SCHELLMAN, 1961; URNES and DOTY, 1961; JIRGENSONS, 1963]. There were other complications. Thus it was found that values for $\lambda_0 = 212$ and for $b_0 = -630$ are correct for completely

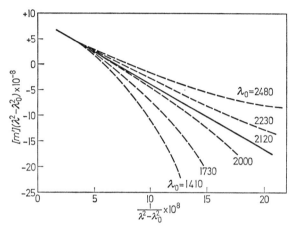

Fig. 7. The rotatory dispersion of poly-γ-benzyl-L-glutamate in ethylene dichloride, plotted by taking various values for the parameter λ_0. (From MOFFITT and YANG, 1956.) λ is expressed in angstrom units

$$\left[\begin{array}{c} -HN-CH-CO- \\ | \\ CH_2 \\ | \\ CH_2 \\ | \\ COOCH_2C_6H_5 \end{array} \right]_n$$

α-helical structures only under certain specified conditions, such as the wavelength range in which the optical rotation was determined (see URNES and DOTY, 1961; URNES, 1963).

Several other semiempirical or empirical relationships have been proposed to express the dependence of the optical activity on wavelength in a meaningful way, *e.g.* by SHECHTER and BLOUT (1964). According to YANG (1965), these do not offer any advantage over the simplified original Moffitt-Yang Eq. (8). Examples showing the calculation of b_0 from the Moffitt-Yang plots will be presented in Chapter V.

8. Configuration and Conformation

As indicated in Chapter I, the optical activity of a protein has *two* causes: the asymmetric carbon atoms and the asymmetry of the polypeptide chain itself. The same is true for any other natural biopolymer, such as a nucleic acid or polysaccharide. The term *configuration* was used formerly for both types of asymmetry. At present, the use of the term *configuration* is restricted to the asymmetry of the carbon atoms linked to four different groups. For the asymmetry of the polypeptide chain backbone itself the term *conformation* is used. A completely disordered polypeptide chain will exhibit optical activity, and this activity is caused by the asymmetric configurations of the constituent amino acids. It is important to note that ordered chains can be optically active even without any asymmetric carbon atoms.

9. The Far Ultraviolet Absorption Maxima and Cotton Effects

All proteins absorb ultraviolet light of wavelengths from about 180 to 240 nm strongly, and the transitions of the resonating chromophoric electrons result in optical activity. The carbonyl groups, $=CO$, of the peptide bonds have such chromophoric electrons which vibrate in an asymmetric environment. The absorption of several polyamino acids and proteins in the 185 to 240 nm spectral zone has been investigated in several laboratories, notably by ROSENHECK and DOTY (1961). All specimens, independent of their conformation and amino acid composition, showed a high absorption maximum at 188 to 195 nm. Weaker maxima as shoulders of the main peak have been observed at the long-wave limb of the absorption curve. It is noteworthy that the absorbance values of all of these maxima are much higher than those of the aromatic side chain absorbances at 270 to 280 nm. Also, it is interesting that the peptide bond absorbance near 190 nm depends only slightly on amino acid composition and amino acid configuration. Rather, the *magnitude* of the absorbance at 190 nm depends on the *conformation* of the polypeptide chain backbone. According to ROSEN-HECK and DOTY (1961), the disordered chain has higher absorbance values than the α-helical chain.

The relationships between the absorbance and optical rotation have been elucidated recently by experimental and theoretical studies. Fig. 8 illustrates this relationship in an idealized form. The solid curves in Figs. 8 a and 8 b represent the changes of optical activity with wavelength near and through the absorption band. The descending part, *d*, of curve 8 a depicts the general behavior of proteins, as was illustrated in Fig. 5. If the measurements are extended to very short wavelengths, the curve exhibits a minimum of levorotation; at still shorter wavelengths, the rotatory power becomes less negative, reaches zero, and then becomes positive. In the example shown in Fig. 8 b, the change of the rotatory power with decreasing wavelength is opposite to that in Fig. 8 a. The absorption curve is shown by the dashed curve. The phenomenon of passage of the rotatory dispersion curve through a minimum (or maximum), zero, and a maximum (or minimum) is called the *Cotton effect,* after its discoverer COTTON (1895). In the case 8 a, when the negative extremum is on the long-wave side, the Cotton effect is *negative;* whereas in case 8 b, when the positive extremum is on the long-wave side, the Cotton effect is *positive.* The proteins, as we shall see later, have several overlapping Cotton effects. A related phenomenon, the *change of circular dichroism* (see Sect. 11) with wavelength near and through the absorption band was also discovered by COTTON (1895). This phenomenon is a measure of the difference between the absorption of the left and right circularly polarized light.

The maximum of a circular dichroism curve is at the same wavelength as the absorption maximum. Circular dichroism may be either positive or negative, whereas absorption is always positive.

In optical rotatory dispersion, the spectral *position* of a Cotton effect is the wavelength at which the optical activity is *zero*, whereas in circular dichroism the position of the Cotton effect is that wavelength at which there is a positive or negative circular dichroism *maximum*.

The quantum theory has been applied to the phenomenon of optical rotatory dispersion by several investigators, and the theoretically developed relation can be written as

$$[m']_\lambda = \frac{96\,\pi\,N}{h\,c} \sum R_i \frac{\lambda_i^2}{\lambda^2 - \lambda_i^2} \tag{9}$$

where the sum represents all optically active electronic transitions involved, λ_i corresponds approximately to the wavelength of each of the absorption maxima, N is the Avogadro number, h Planck's constant, and c the velocity of light. R_i is called the *rotational strength,* and it is related to the previously mentioned factors and to the amplitude of the Cotton effect, A_i. According to URRY and EYRING (1966), the rotational strength can be expressed as

$$R_i = (h\,c/96\,\pi\,N)\,(A_i G_i/\lambda_i) \tag{10}$$

in which G_i is the so called damping factor. The magnitudes of A_i, G_i, and λ_i can be obtained experimentally from the absorbance, circular dichroism, and rotatory dispersion data of the Cotton effect, as shown in Fig. 8 b. According to Eq. (10), the

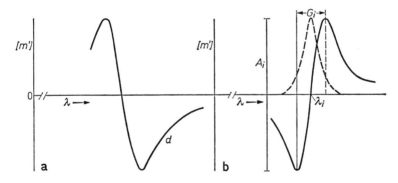

Fig. 8 a and b. Negative (a) and positive (b) Cotton effects. Relationship of the Cotton effect to the absorption curve (dashed curve in b). A_i is the amplitude of the Cotton effect, λ_i is the inflection point wavelength, and G_i the "damping factor" or half band width. (URRY and EYRING, 1966)

rotational strength, R_i, is proportional to the amplitude of the Cotton effect, A_i. If several Cotton effect curves overlap, as is the case in the proteins, the experimental assessment of R_i may be difficult. In the quantum theory of optical activity, the rotational strength is treated as a scalar product of two vectors representing the electric and magnetic moments induced by light of a specified frequency, $\nu = 1/\lambda$.

Since N, h, and c are constants, the expression $96\,\pi N/hc$ of Eq. (9) can be replaced by a constant K. By combining (9) with (3), one gets:

$$[m']_\lambda = [\alpha]_\lambda\, 3\, M/100\, (n_\lambda^2+2) = K \sum_i \lambda_i^2 R_i/(\lambda^2-\lambda_i^2) \qquad (11)$$

which is similar to the Drude equation (4).

The rotatory (or rotational) strength R_i of a transition i can be calculated from molar ellipticity values (see Sect. 11) obtained in circular dichroism measurements. According to MOSCOWITZ (1960):

$$R_i = 0.696\times10^{-42}\times\pi^{1/2}[\theta]_i^0\times\Delta_i^0/\lambda_i^0 \quad \text{(in c.g.s. units)},$$

in which $[\theta]_i^0$ is the ellipticity at the band maximum, and Δ_i^0 is the wavelength interval between λ_i^0 and the wavelength at which $[\theta]_i^0$ falls to $[\theta]_i^0/e$, assuming Gaussian curves. The value of Δ_i^0 is equivalent to the half-width of the absorption band, or measured circular dichroism band, or the "damping factor" G_i (see Eq. 10 and Fig. 7 b). Numerical values of the mean residual specific rotation $[m']_i$, according to MOSCOWITZ (1960), can be calculated from R_i as follows:

$$[m']_i = \frac{R_i}{0.696\times10^{-42}} \times \frac{2}{\pi} \times \frac{(\lambda_i^0)^2}{\lambda^2-(\lambda_i^0)^2}.$$

Again, the last term represents the familiar Drude equation. The calculations to obtain a full rotatory dispersion curve are practical only with the aid of electronic computers. For poly-α-L-glutamic acid and some of its copolymers such computations have been performed by HOLZWARTH and DOTY (1965), and the agreement between the computed and observed curves was quite satisfactory.

The theoretical studies develop in two major directions: one is concerned with the energetics of the stability of the various atomic configurations in the macromolecules (see e.g. FASMAN, 1967) and the other with the interactions of the polarized monochromatic light with the electronic configurations in the atomic structures (see e.g. CALDWELL and EYRING, 1964; TINOCO, 1965; PYSH, 1966; ROSENHECK and SOMMER, 1967). Although some progress has been made in both directions, especially with the relatively simple polyamino acids, there is no way at present to predict exactly what conformation will arise in a chain having a certain amino acid sequence. The problem is very complex because of the many different side chains (R-groups), such as the disulfide bonds, which act as chromophores (IIZUKA and YANG, 1964; COLEMAN and BLOUT, 1968). An even more distant goal is to predict exactly what sort of rotatory dispersion or circular dichroism curve will arise from a certain amino acid sequence in a macromolecule. Thus far, the only evidence of the structural details in the macromolecules of proteins is the experimental evidence provided by the X-ray diffraction studies on crystals. And these have revealed a bewildering complexity of conformations (e.g. PHILLIPS, 1968). A fair agreement between the theoretically estimated and observed rotatory strengths has been achieved only with such relatively simple cases as the helical poly-α-L-glutamic acid (HOLZWARTH and DOTY, 1965).

10. Absorption and Refraction

As indicated in Sects. 1, 2, and 4, the refractive index, n_λ, depends on the wavelength of light. While the optical rotation may be positive or negative, the refractive index is always positive, and it increases rapidly with decreasing wavelength of the light. Within the absorption band, a complex change of the refractive index, a sort of Cotton effect, takes place, as shown in Fig. 9 (see HELLER, 1960).

In the case of protein solutions, one has to consider the refractive index of the solvent. Outside the absorption band, the dependence of the refractive index on wavelength can be approximated as

$$n_\lambda^2 = 1 + a\, \lambda^2/(\lambda^2 - \lambda_a^2)\,, \tag{12}$$

in which a and λ_a are constants. To determine these constants, refractive index measurements are required at least at two wavelengths. The correction $3/(n_\lambda^2 + 2)$

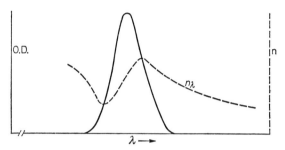

Fig. 9. Change of refractive index n_λ with wavelength near and through an absorption band

becomes important at the short wavelengths, especially for solvents which have high refractive index values. The magnitude of the factor $3/(n_\lambda^2 + 2)$ for water at $\lambda = 589$ nm is 0.794, whereas at $\lambda = 199$ nm it is 0.744. These values are practically the same when diluted buffer solutions are used instead of water. However, for solvents having high refractive index values the factor $3/(n_\lambda^2 + 2)$ is smaller; for example, for 2-chloroethanol at $\lambda = 313$ nm it is 0.719; for $8\,M$ urea solution in water it is 0.739 at the same wavelength.

A selection of the refractive index values, including the factors $3/(n_\lambda^2 + 2)$, has been tabulated by FASMAN (1963).

11. Circular Dichroism

A plane polarized beam of light can be considered as being composed of right and left circularly polarized components (see Sect. 2). In an anisotropic crystal or optically active solution the propagation of the wave components may be different, and the magnitude of this difference depends on both the wavelength and structural properties of the medium. When the frequency of the polarized beam corresponds to the frequencies of the electronic transitions of the chromophore, i.e. within an absorption band of the optically active chromophore, the right and left circularly polarized components are absorbed to a different degree. The phenomenon is called circular dichroism. With the aid of modern instrumentation it now can be measured conveniently, even in the far-ultraviolet spectral zone (see Sect. 5 in Chapter III). As indicated in Sect. 9 of this chapter, circular dichroism is closely related to absorption, and often the circular dichroism maximum coincides with an absorption maximum. However, the magnitude of the dichroism cannot be obtained from molar extinction data, because other factors are involved in addition. When light is absorbed, the electrons are lifted from their ground levels to higher energy orbits; and, when they return to the ground states, the energy is dissipated in the form of heat. These events are not necessarily associated with circular dichroism, since the latter requires definite asymmetry conditions. In some absorption bands the circular dichroism may be negligible, whereas in some other weak bands the circular dichroism and optical activity effects may be strong.

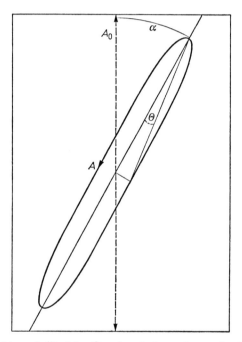

Fig. 10. Graphic definition of ellipticity Θ and optical rotation α of polarized light (TINOCO, 1965). The dashed line A_0 represents the plane polarized monochromatic ray striking the sample. The transmitted ray A is elliptically polarized

The optical density difference between the left and right components, $\Delta D = D_l - D_r$, is measured directly, and, like the rotatory power, ΔD depends on concentration and on the length of the optical path through the sample. This leads to the *circular dichroic absorption* (VELLUZ, LEGRAND, and GROSJEAN, 1965) or *molar circular dichroism*

$$\Delta\varepsilon = \Delta D / c\, d\,,$$

where c is the concentration in gram-moles per liter and d is the length of the optical path through the solution in centimeters. Another way to specify the dichroic absorption is to reduce it to the concentration of 1 g/100 ml and an optical path of 1 cm. This manner of expressing circular dichroism is often used for proteins, and the reduced dichroism is symbolized as $\Delta E^1_{1cm} \%$ or $\Delta D^1_{1cm} \%$.

L Circular dichroism often is expressed also as *ellipticity*, now usually denoted by θ. When a plane-polarized monochromatic ray passes through a circularly dichroic substance, it emerges elliptically polarized, as illustrated schematically in Fig. 10. The ray is directed toward the observer (*i.e.* it is perpendicular to the page); the dashed line A_0 represents the plane polarized incident ray, and the transmitted ray A is elliptically polarized. At the same time, the plane of polarization is rotated (solid line). The tangent of θ is equal to the ratio of the minor to the major axis of the ellipse. The ellipticity, like circular dichroism and optical activity, depends on the concentration and optical path; and the *molar ellipticity*, denoted by $[\theta]$, is expressed in units of degrees$\times$cm^2/decimoles. The *molar ellipticity is directly proportional to the molar circular dichroism*, and it can be shown that

$$[\theta] = 2.303 \frac{4500}{\pi} \Delta\varepsilon = 3298 \, \Delta\varepsilon \,,$$

which is usually rounded up to

$$[\theta] = 3300 \, \Delta\varepsilon \,.$$

The dependence of molar circular dichroism or molar ellipticity on the wavelength of monochromatic light is usually studied disregarding the refractive indices of the optically inactive solvents. This is legitimate when the refractive indices do not differ strongly. For very accurate comparisons, however, the factor $3/(n^2 + 2)$ should be considered (see Sect. 10).

As it was indicated in Sect. 9, circular dichroism and optical rotation are closely related phenomena. Mathematical instruments of general validity, the Kronig-Kramers theorem, are available for the calculation of rotatory dispersion curves from circular dichroism (ellipticity) data and *vice versa*. These calculations have been reviewed in some detail by MOSCOWITZ (1960), and the subject has been treated in a general fashion by MOFFITT and MOSCOWITZ (1959), and by CALDWELL and EYRING (1964), among others. The applications to polypeptides and proteins have been considered, *e.g.* by TINOCO (1965), HOLZWARTH and DOTY (1965), CARVER, SHECHTER, and BLOUT (1966), BEYCHOK (1966, 1967), and TINOCO and CANTOR (1970).

The interaction of light with the electronic configurations of asymmetric substances is studied by measuring either the rotatory dispersion (ORD) or circular dichroism (CD) effects. Since the rotatory dispersion effects are caused by refractive phenomena, *i.e.* one of the circularly polarized components is slowed down more than the other (see Sect. 2), these effects can be measured at any wavelength. Circular dichroism, on the contrary, can be detected only in the wavelength region of the optically active absorption band, *i.e.* at frequencies corresponding to the dominant frequencies of electronic transitions from normal to excited states. In a general fashion, the relationship between the Cotton effects in rotatory dispersion and circular dichroism of proteins with a significant α-helix content is shown in Fig. 11. The major absorption bands responsible for the observed optical activity effects are centered at 190—192 nm, 205—210 nm, and at 222 nm.

The Kronig-Kramers relationship can be written in the form

$$[M]_\lambda = 2/\pi \int_0^\infty [\theta]_{\lambda_1} \left(\frac{\lambda_1}{\lambda^2 - \lambda_1^2} \right) d\lambda_1 \,,$$

in which $[M]_\lambda$ is the molar rotation at wavelength λ and $[\Theta]_{\lambda_1}$ is the molar ellipticity at wavelength λ_1. Calculation of the $[M]_\lambda$ values is usually done with the aid of computer programs at wavelength intervals of 1 nm. The integration, of course, cannot be done from 0 to infinity but only within the optically active bands whose CD was measured. The infrared and longer-wave effects, if any, can be neglected, but optically active bands are suspected in the very deep ultraviolet zone at wavelengths below 180 nm. For example, STURTEVANT and TSONG (1969) calculated the molar rotations from measured molar ellipticities for cytochrome-b_2 in the accessible spectrum and found discrepancies between the calculated and observed rotatory dispersion values. An analysis of these discrepancies indicated the possible presence of optically active bands in the vicinity of 150 nm.

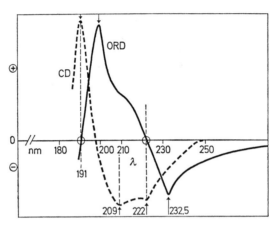

Fig. 11. Relationship between circular dichroism (CD) and rotatory dispersion (ORD) of a protein with a significant α-helix content. The positive CD band maximum at 191 nm corresponds to the crossover point of the ORD curve at 191 nm and is close to the peptide bond absorption maximum. Similarly, the negative CD band maximum at 222 nm corresponds to the crossover point (zero rotation) of the ORD at 222 nm. The CD curve approaches the base line near 250 nm, and the CD=0 at 250 nm and 260 nm. The ORD curve, however, remains in the negative part of the chart. The measured band maximum values depend on the helix content

The negative CD band of helices centered at 222 nm (Fig. 11) is caused by the relatively weak $n - \pi^*$ transition of the peptide bond $- CO - NH -$ (SCHELLMAN and ORIEL, 1962). The mechanism of interaction can be considered as limited to the $2p_y$ orbital on the oxygen. In the interaction with light, the electron is transferred from a nonbonding orbital to an antibonding π orbital. This transition involves a circular motion of the electric charge-inciting magnetic moment directed along the $C - O$ bond, and the encounter, of course, affects the light wave too. The final effect from the interaction of the wave with the polypeptide depends not only on the electronic structure of the peptide chromophore but also on the *order* of these chromophores and on the *vicinal effects* of other atomic groupings. (In other optically active carbonyl group-containing compounds the Cotton effect of this transition is found centered at different wavelengths and can be positive; thus camphorsulfonic acid displays a positive effect near 290 nm.)

The strong positive CD band centered at 191 nm and the negative peak at 205—210 nm (Fig. 11) are caused by the $\pi - \pi^*$ transition. This interaction involves higher absorption than in the $n - \pi^*$ transition. The polarization of the electric transition dipole moment in the $n - \pi^*$ transition is perpendicular to the plane of the peptide chromophore, whereas in the stronger $\pi - \pi^*$ transition it is in the plane of the peptide group and there is a change in the charge distribution along the $C-O$ bond. The appearance of two Cotton effects of opposite sign in the 190—210 nm spectral zone is explained as "exciton splitting". It has been shown that the resonance interaction of the identical peptide chromophores along and across the axis of the helix should lead to the splitting of the peptide absorption band, and this theoretical prediction has been verified experimentally. A more detailed treatment on the far-ultraviolet absorption spectra and origin of the CD bands can be found, e.g. in the reports of GRATZER (1967), WOODY and TINOCO (1967), and TINOCO and CANTOR (1970).

Beside the described polypeptide "backbone" chromophores, there are chromophoric groups of atoms in the side chains. The disulfide bonds of the cystine residues are inherently asymmetric and are known to produce CD bands at 195—200 nm and at 250—260 nm (COLEMAN and BLOUT, 1968). Moreover, the R-groups of tryptophan, tyrosine, and phenylalanine display relatively weak CD bands in the 230—310 nm spectral zone. However, while the $-S-S-$ bonds are inherently asymmetric, the aromatic residues are not; i.e. their chromophoric effects depend on the asymmetric microenvironment in the macromolecule. Depending on these vicinal effects, the CD bands of the same chromophore (e.g. tyrosine) may be positive or negative, the peaks can be found at different wavelengths and differ in rotatory strength. Although the tryptophan bands are centered mostly in the 290—310 nm region (FRETTO and STRICKLAND, 1971; IKEDA et al., 1972), some of them extend to the shorter wavelengths and overlap with the tyrosine and phenylalanine effects. In the 250—260 nm zone, the aromatic group effects overlap also with the disulfide effects. For all these reasons, a detailed assignment of the bands of the very complex CD spectra of proteins is difficult. But in spite of these difficulties, the study of these complex spectra is much more promising than the observation of rotatory dispersion, because the ORD curves in this spectral zone are either smooth or have only a few humps. These CD bands in the near-ultraviolet zone are more sensitive to small conformational changes than those in the 190—225 nm region or than the ORD effects. Some examples will be treated in Chapter VII.

The chromophoric effects of the R-groups may be displayed even in the far-ultraviolet zone, e.g. a tyrosine band has been observed in the CD spectra of immunoglobulins at 232—235 nm (CATHOU, KULCZYCKI, and HABER, 1968; DORRINGTON and SMITH, 1972). In proteins of moderate or high helix content (30—70%), however, the polypeptide backbone effects completely overshadow the side group effects in the far-ultraviolet. An important factor in these considerations is simply the *large number* of the peptide bonds in comparison to one or another R-group. In ordered structures this effect of number is compounded by the effects of *order*.

12. Magnetic Circular Dichroism

Magnetic optical activity, discovered more than 100 years ago by M. FARADAY, is known as the Faraday effect. He found that optical activity can be induced in any substance exposed to a magnetic field of sufficient strength. While in natural optical activity (ORD and CD) the measured effects depend on the asymmetric structure of the material, in magnetic optical activity the effect depends on the strength and asymmetry of the external magnetic field. The quantitative relationships were studied in detail by VERDET (1854—63) who found that, at constant wavelength, the rotation of the plane of polarization α_M is proportional to the length of the light path through the substance (l), the strength of the magnetic field (H), and the cosine of the angle between the direction of the ray and magnetic force lines:

$$\alpha_M = \delta \cdot l \cdot H \cdot \cos \varphi,$$

in which δ is a proportionality constant (Verdet's constant) characteristic for each substance. If the ray and the magnetic force lines are parallel, as can usually be arranged, $\cos \varphi = 1$, and $\alpha_M = \delta \cdot l \cdot H$. In the case of a solution, the measured effect is the sum of the effects of the solvent and the solute, i.e. $\alpha_M = \alpha_{M\,solvent} + \alpha_{M\,solute}$. The specific magnetic rotatory power, $[\alpha_M]_\lambda$, is defined as $[\alpha_M]_\lambda = \alpha_M / c \cdot l \cdot H$, in which c is the concentration of the solute in g/ml. In these relationships the light, including its wave-

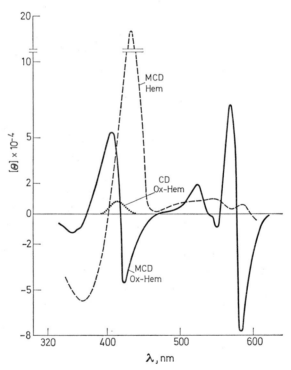

Fig. 12. Magnetic circular dichroism (MCD) of hemoglobin (broken curve) and oxyhemoglobin (continuous curve) in a strong magnetic field of 49,600 gauss. The effects are expressed in terms of mean residual molar ellipticities on a strongly compressed ordinate. On the same scale, the effects of the natural CD (circles) near the 420 nm region appear much weaker than those of the MCD. The proteins were dissolved in 0.1 M phosphate buffer, pH 7.0. (After DJERASSI, BUNNENBERG, and ELDER, 1971.)

length, is constant. However, as in natural ORD and CD, the magnetically induced effects can be measured with light of different wavelengths, keeping c, l, and H constant. This yields *magnetic optical rotatory dispersion* (MORD) if one uses plane polarized light of various wavelengths and *magnetic circular dichroism* (MCD) if the plane polarized light is modulated in the right and left circularly polarized components.

Although the magnetic field effects have been extensively utilized in modern instrumentation for optical activity (see next chapter), magnetic rotatory dispersion has found only a very limited use in structural studies. More attention recently has been paid to the magnetic circular dichroism (see *e.g.* DJERASSI, BUNNENBERG, and ELDER, 1971). In this approach, it is important that MCD can be tremendously enhanced by using a very strong magnetic field. Thus the method is of great analytical value for detecting trace amounts of substances having characteristic MCD spectra. For instance, the method was used in an attempt to detect traces of porphyrins in moon dust; nothing was found and the absence of porphyrins was regarded as proof that living matter was unlikely to be present on the moon. The MCD spectra of proteins have more complex fine structure than those of the natural CD, but there is little progress in the interpretation of the MCD bands. A critical report on the application of MCD to the study of protein conformation has been provided by BARTH et al. (1972). An example of MCD of hemoglobin and oxyhemoglobin in a very strong magnetic field is illustrated in Fig. 12. Note the striking difference between the magnitude of the CD versus MCD at 400—440 nm.

Chapter III

Polarimeters and Spectropolarimeters.
The Measurement of Optical Activity

Ordinary light coming from the sun or a lamp is *polychromatic, i.e.* of *different wavelengths,* and the waves oscillate in *different planes* perpendicular to the direction of propagation. A *monochromator* separates the waves of certain definite length from the polychromatic multitude. A *polarizer* separates out the waves which oscillate only in one plane. A monochromatic unpolarized ray can be compared with a bottle brush in which the bristles point out in various directions, and a polarizer works like a slicer cutting off the bristles at the sides, leaving only those which point "up" and "down". Since the uniformity of the light is achieved not by "transforming" but by *separation,* the polarized beam is always of lesser intensity than that generated by a source.

1. Visual Polarimetry

Simple visual polarimeters have been used for more than a century, and their main parts are: a light source, a set of polarizer and analyzer prisms, a sample cell or tube, and a circular scale graduated in degrees and fractions of degrees. There are various types of prisms and polarimeters, but the principle is the same, as illustrated schematically in Fig. 13. In this

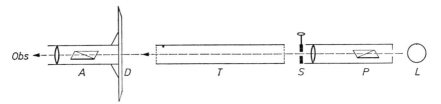

Fig. 13. Schematic drawing of a simple polarimeter. *L* Light source, *P* Polarizer, *S* Slit, *T* Sample tube, *D* A graduated disc rotating with the analyzer *A*

arrangement, the prisms are built of specially cut and combined large calcite crystals (Nicol prisms) which exhibit double refraction, and in which the ordinary ray is separated from the extraordinary ray. One of the prisms serves as the *polarizer (P)* and the other identical unit is used as the *analyzer (A).* The emerging light has its maximum intensity when the principal planes of P and A are parallel.

If P remains fixed and the analyzer is rotated, the intensity of the transmitted light decreases and becomes minimal when the principal planes are at right angles. If the analyzer is rotated further in the same direction, maximum brightness is restored after a further 90° rotation. The measurements are performed by zeroing the empty instrument (or the tube with the optically inactive solvent) at minimum transmission; the sample then is inserted between the polarizer and analyzer and the effect is observed through the eyepiece. If the specimen is optically active, the change in brightness, corresponding to the optical activity, is measured by rotating the analyzer until the original condition of the field of vision is restored. The required angular degree, or fraction of a degree, α, is the rotatory power. In the most common optical arrangement, the circular field of vision is divided in half by a hair line, and the magnitude of α is determined by adjusting the analyzer so as to equalize the brightness of the halves. The rotation may be either positive (clockwise) or negative.

Sodium light in which the principal wavelength is the yellow line of the sodium spectrum at 589 nm is *most* often used in visual polarimetry. Formerly, a circular platinum trough containing sodium chloride was placed in the hottest section of the Bunsen flame, while now the more convenient sodium lamps are used. Mercury lamps emitting chiefly the wavelengths of about 436, 546, and 579 nm are occasionally used, and a partial separation of the colors is possible by using appropriate light filters. A detailed description of visual polarimeters and their components can be found in the book of LOWRY (1935) and in the review by HELLER (1960).

2. Photoelectric Spectropolarimetry

Visual polarimetry is of very limited value for the study of the conformation of macromolecules. To employ the most effective ultraviolet light, one has to design polarimeters capable of detecting the invisible rays. This is possible in various ways, for instance with the aid of photography. All modern spectropolarimeters are equipped with the very convenient photoelectric devices for signal detection.

The principal parts of a photoelectric spectropolarimeter are: (1) a *light source* yielding sufficiently intense ultraviolet radiation, (2) a *monochromator*, (3) a *polarizeranalyzer* prism system, and (4) a sensitive *photoelectric detector* device. The recording instruments, in addition, have a *recorder* which plots the optical activity against wavelength. Some of the more sophisticated modern instruments even plot other essential data, such as the absorbance, slit width, and the phototube activating voltage.

Modern ultraviolet spectropolarimetry has an interesting history. In the first attempts, in 1950 to 1955, only the near ultraviolet spectral zone could be reached (to about 300 nm). The main difficulties were not in detecting the signal or ability to record the result but in the *low intensity* of the ultraviolet rays and their *high absorbance* in the optical system and samples. Problems were also encountered in obtaining a sufficiently *monochromatic* light, in *stability* of the light sources, and in precise determination of the *extinction point*. The last problem was the first to be solved, by adapting the method of *symmetrical angles* (RUDOLPH, 1955). The angle of rotation was not determined directly by finding the minimum deflection of the photo-meter, but by taking two analyzer readings at equal galvanometer deflections above and below the extinction, and averaging the two. Even higher precision was achieved by making a whole series of readings, averaging the positive and negative readings, and finding the extinction by extrapolation. Later, the manual settings were eliminated by introducing a motor-driven device which oscillates the prism continuously at a certain preset symmetrical angle. Fig. 14 shows such an instrument.

One of the major obstacles in extending the measurements into the far ultraviolet was the high absorbance of ultraviolet light in the various materials used for building the lenses, prisms, and tube windows. Table 1 lists the approximate wavelength limits for some of these materials.

According to these data, glass and calcite are useless for working in the far ultraviolet. Quartz is the most important material to be used for lamp envelopes, lenses, prisms, and sample cells. *Crystalline quartz* is used for the polarizer-analyzer system, whereas *amorphous fused quartz* (silica) is used for lamps, lenses, and sample cells. Since natural quartz crystals often contain strongly absorbing impurities, the polarizer and analyzer prisms are now made of crystals artificially grown from the purest starting materials. The same purity requirements, of course, apply for the fused silica used for the other optical components. With good quartz

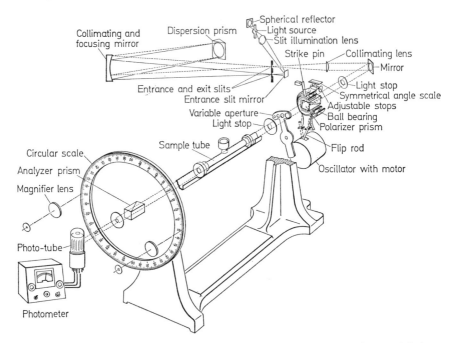

Fig. 14. A simple photoelectric spectropolarimeter. The dispersion prism is the essential element of the monochromator

Table 1. Approximate limits of useful light transmission through various substances
(after HELLER, 1960)

Substance	Approximate shortest λ, nm
Glass discs, ordinary, 2 to 3 mm thick	365
Special glass, high transmittance, 10 mm thick	310—330
Film of Canada balsam in Nicol prisms	280—340
Calcite, 4 cm layer	240
Quartz, 2 cm layer	185
Fluorite, pure crystals 2 cm layer	100

optics it is possible now to reach about 180 nm without much difficulty. However, factors other than the transmittance of the optical components then become important, such as the *absorbance by the sample* and by the *oxygen* of the air. The absorbance by the sample is reduced by working with *diluted solutions in thin layers,* and the oxygen effects by air-tight enclosure of the system and flushing it with nitrogen. The oxygen effects are noticeable at the wavelength of 195 nm and below.

The *light source* has always been very important in far-ultraviolet spectropolarimetry. The major problem is the energy distribution as a function of wavelength, most of the energy emitted having wavelengths of 300 to 600 nm. The high-frequency (or very short wave) rays are of very low intensity, but the total intensity can be raised by increasing the wattage. Regarding the general type of emission, two types of lamps must be distinguished: the lamps producing a *line spectrum,* and the *continuous emission* sources. *Mercury lamps* are used as the most suitable sources of strong line spectral emission. Mercury has emission maxima of the wavelengths: 578 to 579, 546 (green), 492, 436 (blue), 405 (violet), 390, 365 to 366, 334, 313, 296.7, 275.2, 265.5, and 253.6 nm. (Since most of the lines have a composite fine structure, several of the wavelength values given above are approximations.) In the far ultraviolet, below 254 nm, there are no strong mercury emission maxima. Mercury lamps are useful for manul spectropolarimetry when the measurements are made at the wavelengths of the spectral lines. They are not recommended for recording spectropolarimetry in the visible and near ultraviolet because of the very uneven energy distribution. *Xenon lamps* are the most important continuous emission sources. The 450 to 500 watt direct-current xenon lamps produced by Osram are now used in practically all recording spectropolarimeters. The Engelhard-Hanovia company (Newark, New Jersey) now also offers higher-wattage xenon and xenon-mercury lamps, e.g. of 1000, 2500, and even 5000 watt engery, but special power supplies are needed for these lamps. The length of the arc in the lamps is 3 to 8 mm, and the radiation intensity is not uniform over the arc. Perfect stability of the arc would be important, but it is not attainable with even the most sophisticated voltage regulation. Direct-current lamps usually are preferred because of greater stability of the arc. Very good results were obtained in our laboratory with alternating-current mercury argon lamps (1000 watt). Lamps of lower energy output, usually in the 150 to 250 watt range, are used extensively for measurements in the visible and near-ultraviolet spectrum. The low pressure Hanovia SH mercury lamps are useful for high-precision measurements at 546.1, 435.8, 404.7 nm, and even for the near ultraviolet. Since the emission maxima of this lamp are very sharp, it is used also for the calibration of monochromators.

The *monochromator,* the device filtering out certain wavelengths from the beam, is one of the most important components of a spectropolarimeter [1]. In the early stages of photoelectric spectropolarimetry (1956 to 1960), the simple single-prism monochromators were found satisfactory. Attempts to employ them also for the far ultraviolet, however, led to unexpected difficulties and errors [URNES and DOTY, 1961]. It appeared that the main difficulty was caused by *stray light.* The intensity of ultraviolet light passing through a protein solution (or any other strongly absorbing system) decreases strongly below the wavelengths of 240 nm. To have some detectable intensity of light passing through the sample, one has to open the monochromator slit wide. However, the wider the slit, the more *inhomogeneous* is the radiation. The short-wave components of the inhomogeneous radiation are absorbed and scattered more than the long waves. The result is that only the long waves reach the detector. The errors may be compounded if the optical elements have absorbing or fluorescent impurities or if the absorbance of the sample is very high. For these reasons, the early publications in far-ultraviolet spectropolarimetry (1961 to 1962) must be considered with caution. Attempts were made to obtain purer light by placing two single-prism monochromators in series. The results of these endeavors varied, depending on the quality of the particular optics and light sources used. The best results by this method in the far ultraviolet were obtained by BLOUT, SCHMIER, and SIMMONS (1962).

After 1962, it became clear that the answer was to be found in *compact double prism monochromators* receiving light from *high-intensity light* sources. The double-prism monochromators are now produced by all major builders of spectropolarimeters for use in Cotton effect studies in the far ultraviolet.

Recording Spectropolarimeters. The first practical commercially available recording instruments were provided by the Rudolph company. They had the single-prism monochromator, and they were useful down to the wavelength of approximately

---

[1] The major part of a monochromator is either a dispersion prism or a grating serving the same purpose. The other parts are: reflectors, adjustable slits and a wavelength scale.

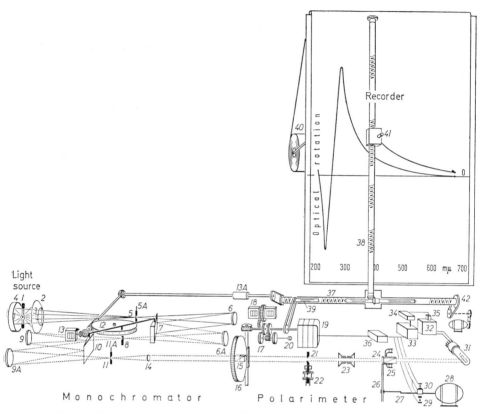

Fig. 15. Schematic drawing of a Rudolph RSP-3 recording spectropolarimeter

1 Light bulb	*11A* Exit slit width adjust and indicator	*28* Oscillator motor
2 Spherical reflector	*12* Wavelength cam	*29* Magnetic chopper
3 Slit illumination lens	*13* Wavelength counter	*30* Chopper switches
4 Focusing reflector	*13A* Monochromator connecting bushing	*31* Photomultiplier
5 Entrance slit of monochromator		*32* Photomultiplier power supply
	14 Collimating lens	
5A Entrance slit width adjust and indicator	*15* Polarizer prism	*33* Pre-amplifier
	16 Servo-driven worm gear	*34* Photomultiplier monitor servo amplifier
6 First collimating reflector		
	17 Servo-driven optical rotation gearing	*35* Photomultipler voltage monitor servo motor
6A First spectrum focusing reflector		
	18 Servo-driven optical rotation counter	*36* Polarizer servo amplifier
7 First dispersion prism		
8 Second or intermediate slit	*19* Polarizer servo motor	*37* Wavelength and time-base lead screw
	20 Optical rotation range selector	
9 Second collimating reflector		*38* Optical rotation lead screw
	21 Variable aperture	
9A Second spectrum focusing reflector	*22* Quartz control plates	*39* Optical rotation spline shaft
	23 Sample cell	
10 Second dispersion prism	*24* Analyzer prism	*40* Chart paper
	25 Rotary bearing	*41* Pen
11 Third or exit slit	*26* Oscillator arm	*42* Wavelength and time-base drive
	27 Oscillator drive shaft	

240 nm. Meantime, in 1961 to 1963, the Applied Physics Division of the Varian Corporation (Monrovia, California) developed the Cary recording instrument which was useful for the study of the Cotton effects. The main features of this instrument were a double-prism monochromator and the use of the Faraday effect for modulation of the polarized ray. A recording instrument similar to that of the Rudolph type was developed in Japan and offered in this country by the Durrum Instrument Corporation (Palo Alto, California). This Durrum-Jasco instrument has a double prism monochromator, uses the oscillating prism principle for finding the extinction point, and has provisions for measuring and recording the absorbance, for regulation and recording of the slit width and phototube activating voltage. In 1963 the Rudolph Instruments Engineering Co. built a few recording instruments which were specially designed for studies in the far ultraviolet; the model RSP-3 instrument has been in use in our laboratory for several years. For protein studies in the 190 to 240 nm wavelengt zone, it has proved equivalent to or better than competitive spectropolarimeters. The optical, electrical, and mechanical systems are illustrated schematically in Fig. 15.

The principle of Cary Model 60 recording spectropolarimeter is shown in Fig. 16. The light source, the 450 to 500 watt xenon lamp, is the same as in the Rudolph instruments. The double prism monochromators also are similar except that the Cary instruments have wider angle (30°) dispersing prisms than the Rudolph instruments.

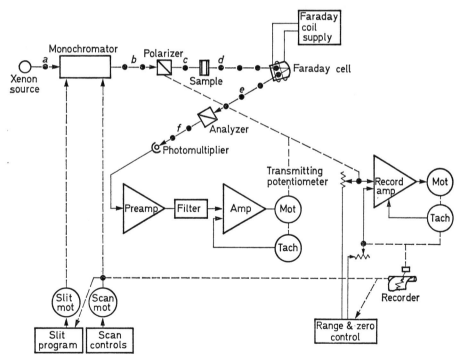

a=Undispersed, non-polarized beam   d=Beam "c" rotated by sample
b=Monochromatic non-polarized beam   e=Beam "d" cyclically displaced by Faraday cell
c=Monochromatic polarized beam   f=Component transmitted by analyzer

Fig. 16. Explanatory chart of the Cary Model 60 recording spectropolarimeter

The angle of rotation in the Cary instruments is found by using the Faraday effect. A Faraday cell consisting of a silica cylinder surrounded by a coil is placed ahead of the analyzer. An alternating current is passed through the coil, cyclically displacing the plane of polarization of the ray. After passing the analyzer, the ray strikes the photomultipler tube, the current is amplified and the signals are transmitted to the polarizer and recorder. The polarizer and analyzer prisms in the Cary instruments are made of ammonium dihydrogen phosphate; the angle of incidence of the refracting surface of the crystals is 60°, and they are immersed in cyclohexane. The instrument has been described in some detail in a publication of CARY et al. (1964). A critical survey of the mentioned spectropolarimeters and others has been contributed by CARROLL and BLEI (1963), and a study on the sensitivity of photoelectric spectropolarimeters has ben published by BÜRER and GÜNTHARD (1960). They conclude that sensitivity depends mainly on the available light intensity, and that half-shade angle, usual optical imperfections in the polarizer-analyzer combination, and the dark current of the photomultipler are less critical.

3. Optical Rotatory Dispersion Measurement.
General Principles of Operation. Sources of Errors.
Accuracy and Reproducibility

The operational conditions for making measurements on a sample depend on the aims of the investigator, the absorptive properties of the substance, and the *available amount of sample*. If one has an ample amount of a colorless protein, one may be content with determining the Moffitt constants; for this purpose, the determinations can be made in the spectral region of 300 to 600 nm. If the available quantity of the specimen is very limited, *e.g.* 0.5 mg, one could start the determinations at a low wavelength (260 or 270 nm) and proceed down the wavelength scale. For the determination of the Moffitt constants (see Chapter II) it is advisable to avoid the aromatic group absorption zone at 275 to 280 nm, because some proteins exhibit weak Cotton effects at these wavelengths.

For the sake of uniformity of experimental conditions, it might seem preferable to run a certain solution of constant concentration in the same cell over the whole attainable spectral range of about 190 to 700 nm. This is impossible because of the *vastly different intensity of the effects in the various ranges*. With visible light, the effects are so weak that to achieve sufficient precision one has to use 0.2 to 1.0% solutions in 1 or 2 decimeter tubes. One cannot proceed with the same solutions and tubes into the far ultraviolet because of the *prohibitive absorbance*. The rotatory dispersion in the near ultraviolet at 250 to 400 nm is determined by using approximately 0.1% solutions in cells having an optical path of 1.0 or 5.0 cm. Since the optical activity effects are much more intense in this zone than in the visible, the precision is not affected. For the work in the *far* ultraviolet Cotton effect zone of 190 to 240 nm, however, even these small amounts of the specimen are too large. For these far-ultraviolet runs the concentrations must be of the order of 0.003 to 0.03%, and the cells should have optical paths of 0.20, 0.10, or 0.05 cm.

The monochromator *slit width* and the phototube *activating voltage* are other important parameters. The wider the slits the less monochromatic is the light, even if the instrument has a double prism monochromator. However, at a too narrow slit the energy entering the sample may be too weak. In the modern recording spectropolarimeters the slit width is programmed or regulated automatically. It is gradually widened from about 0.1 mm in the visible and near ultraviolet to about 2.0 mm in the far ultraviolet. The extremely high absorbance of the proteins at 190 to 220 nm is the major cause of difficulty. Experience has shown that the permissible maximum absorbance is approximately 2.0, and that the values of 1.5 to 1.6 represent safer limits than the maximum value. It is preferable to keep the phototube activating voltage constant, but for the sake of sufficient response of the detector it may be necessary to increase the voltage in the far ultraviolet. The use of 400 to 1200 volts is common for this purpose. A very high activating voltage results in high electrical noise.

Table 2. Optical density (absorbance) of 0.1% solutions of proteins at 278 to 280 nm, and an optical path of 1.0 cm. (Compiled from many sources including our own determinations)

Alcohol dehydrogenase, liver	0.42
Alcohol dehydrogenase, yeast	1.26
Aldolase, muscle	0.91
α-Amylase, B. subtilis	2.56
α-Amylase, pancreatic	2.50
Bovine serum mercaptalbumin	0.66
β-Casein	0.46
Chymotrypsinogen	2.0
Deoxyribonuclease	1.15
Elastase	1.85
Enolase	0.89
γ-Globulins (immunoglobulins)	1.3—1.5
Glutamic dehydrogenase	0.97
Glyceraldehyde-3-phosphate dehydrogenase	1.0
α-Glycerophosphate dehydrogenase	0.75
Human serum mercaptalbumin	0.53
Latic dehydrogenase	1.40
β-Lactoglobulin	0.97
Lysozyme (muramidase)	2.64
Myoglobin, sperm whale	1.87
Papain	2.49
Pyruvate kinase	0.54
Ribonuclease	0.71
Soybean trypsin inhibitor	1.06
Trypsin	1.56

Complete compilations of absorbance data of proteins have been provided by Kirschenbaum (1970, 1972).

The attainable *degree of precision* in rotatory dispersion measurements depends on many factors aside from the quality of instrumentation. Of dominating importance are: the *absorbance* of the sample and the *magnitude of the effect of optical activity*. Low absorbance and high optical activity are favorable for high precision. Poly-α-L-glutamic acid, poly-α-L-lysine, and serum albumin are favorable cases; ribonuclease is less easy to work with, especially at wavelengths below 300 nm. Even greater difficulties are encountered in work with the immunoglobulins, if one wants to study their rotatory dispersion in the far ultraviolet. The absorbance values of a group proteins at 280 nm are presented in Table 2.

The smallest differences which can be measured in angular degrees are about ±0.001°. However, such precision can be attained only under the most favorable conditions with the best instruments. A value of ±0.002 or 0.003° is more realistic. The possible percentage error, then, depends chiefly on the rotatory effect at the particular wavelength. If the precision is ±0.002° the probable error will be ±1% for an optical activity value, α, of 0.20°, but it will be ±10% if the angle is only 0.02°. The angles measured in the far ultraviolet amount to about 0.02 to 0.4°, thus the precision varies considerably. In those cases where the absorption is high and the

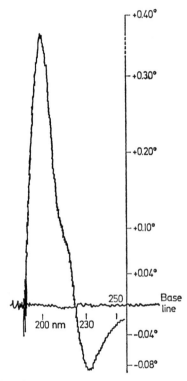

Fig. 17. A tracing of the far ultraviolet rotatory dispersion curve of a diluted solution of poly-α-L-glutamic acid

rotatory effect is small, additional errors and difficulties are caused by the electrical noise, *i.e.* low *signal-to-noise ratio*. Figs. 17 to 19 show tracings from runs in the far ultraviolet. In Fig. 17 we have a favorable case, a 0.09% solution of poly-α-L-glutamic acid in aqueous solution of pH 4.9 recorded in a cell with an optical path of 0.10 cm. The two overlapping Cotton effects, a negative one with a trough at 233 nm and a positive one with a peak at 198 nm, are clearly outlined. The optical activity through most of the curve is high, the electrical noise is insignificant. In Figs. 18 and 19 we have a less favorable case: ribonuclease-A in aqueous solution. Because of the high absorbance and weak effects, a 0.02% solution, in a cell with an optical path of 0.50 cm, had to be used at 210 to 250 nm. The two parallel recordings illustrate the reproducibility. Fig. 19 shows the recordings near 200 nm; the reproducibility is poorer and the electrical noise greater than in the runs at longer wavelengths. It is advisable to run the recorder at a *low scanning speed* and *repeat the runs* 3 to 5 times. Generally a curve then is estimated from the arithmetic average of the runs.

Optical artifacts due to stray light and other disturbances are always a hazard when one works with strongly absorbing substances in the far ultraviolet. A reasonably good reproducibility is not a sufficient guarantee that no artifacts are involved. A good procedure to ascertain the absence of such spurious effects is *to repeat the runs*

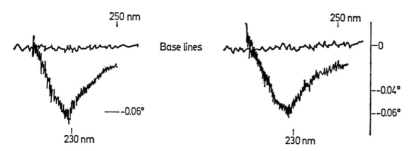

Fig. 18. Tracings showing reproducibility of the curves in the far ultraviolet. 0.02% ribonuclease in aqueous solution; optical path 0.50 cm; slow scanning speed

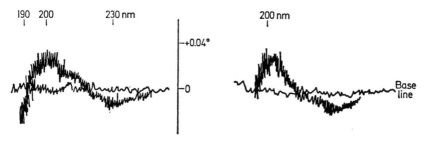

Fig. 19. Tracings showing reproducibility at 190 to 240 nm. Repeated runs of an 0.02% aqueous solution of ribonuclease; optical path 0.10 cm; slow scanning speed

at different optical paths or different concentrations. In the absence of artifacts, the *optical activity*, α, at any wavelength *should be proportional to the optical path and concentration.*

Fig. 20 shows the strong increase in the absorption of a protein as one approaches the 190 nm region. A tracing of a recorded rotatory dispersion curve is also shown. The left ordinate indicates the optical density and the right ordinate the optical activity in terms of specific rotation. In this case (a diluted solution of immunoglobulin) the rotatory effects are very weak and the errors are large.

Significant changes in the optical activity of some proteins were observed by WILSON and FOSTER (1972) after long exposure of the solutions to ultraviolet light of wavelengths below 240 nm, and the changes were attributed to *photochemical destruction* of the protein. Photolysis was noticeable in proteins of high helix content, *e.g.*

serum albumin, but it was observed also in some nonhelical proteins at 50—60°. Such photochemical effects can be verified by comparing measurements made on aliquots exposed for long periods to the light with those for unexposed aliquots. If a systematic change is noticed in repeated recordings of a sample, the photochemical effect should be checked, especially when the experiment is made at elevated temperatures. (No clearly reproducible photolysis effects have been observed in our laboratory on many proteins at 22—25°.) With respect to the development of in-

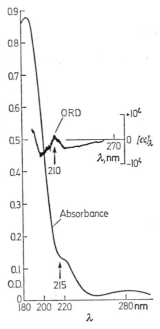

Fig. 20. Rotatory dispersion and absorbance of γ-globulin (human immunoglobulin IgG) in the far ultraviolet. Optical density is plotted on the left ordinate, specific rotation in angular degrees is plotted on the right ordinate. [JIRGENSONS, 1966 (3)]

strumentation, the work of Wilson and Foster shows that the right approach in enhancing weak optical activity effects in biological macromolecules is not to increase the intensity of light but rather to improve the electronic amplification and noise elimination.

4. Cells and Solutions

The choice of tube or sample cell is of the utmost importance in polarimetry. The older type polarimeters have sample compartments holding 40 cm long tubes, although 10 or 20 cm tubes are mostly used. The modern spectropolarimeters have small sample compartments, because the long tubes are not adequate for work in the ultraviolet zone. Cells with short optical paths of between 0.01 to 10 cm are now used for structural studies; and work on the ultraviolet Cotton effects requires the shortest path cells of 0.01 to 0.50 cm. The most important factor in the choice of a cell is the optics of the cell windows. Two major types of cells are now used: *cells with removable windows* which are tightened by means of screw caps, and *cells with fixed windows*. The cells with removable windows are used chiefly for measurements in the visible and near ultraviolet. They are easy to clean and to fill. If only small volumes of a solution are available, semimicro tubes with narrow bore can be used. However, several hazards have to be considered in using these tubes, such as reflection from the walls of the tube and stress birefringence caused by tightening the window plates. The short path screw cap cells are now made of teflon, and plastic washers are used for better sealing. The window discs are made of birefringence-free fused silica. If these cells are used, it is advisable to test the empty tightened cell for optical activity, because of the possible stress effects. The very short path 0.1 and 0.2 cm screw cap cells are not recommended for another reason—the uncertainty of calibration regarding the precise length of the optical path. Slight differences in tightening the windows may result in a significant difference in the length of the optical path. If these cells are calibrated by determining the absorbance or optical activity of known solutions, the windows should not be removed. For these reasons, fixed-window cells are preferred. They are easy to calibrate, fill, and wash. The wide-aperture cylindrical short-path cells provided by the Pyrocell Manufacturing Co. (Westwood, New Jersey) have been found very satisfactory for the work in the far ultraviolet zone.

Measurements in the far ultraviolet are not affected by the color of the solution. Turbidity, however, should be avoided, as it may lead to uncontrollable effects of *depolarization* and scattering. Slight aggregation could possibly cause a change in optical activity, depending on the system. Special precautions must be observed with solutions of unstable solutes, such as some enzymes, if long, repeated series of measurements are made at room temperature. The cell and sample compartment should then be kept at a suitable lower temperature by cooling. Dew formation on the windows of the cell, however, must be avoided since this has the same disturbing effects on the ray as turbidity or dust in the way of the light.

Selection of buffers and solvents of *low absorbance* is another important consideration when planning measurements in the far ultraviolet. Phosphate buffers should be preferred, and carbonates, acetates, and any other substances containing the carbonyl group should be avoided, because of high absorbance.

Accurate solution *concentration* data are as important as calibration of the thin layer cells. If the substance is available in large amounts, several analytical procedures can be used for concentration determination. If, however, only a few milligrams are at hand, the concentration determination is a formidable problem. Protein concentration usually is determined either by measuring the optical density at 280 nm (Table 2) or by micro-Kjeldahl nitrogen analysis. If these data are not available, the safest way is to determine the concentration by *gravimetric microanalysis, i.e.* by weighing the dried sample to a precision of 0.01 mg. One has to be sure then that the weighed sample is free of moisture and other extraneous material, such as buffer. The latter can be removed by dialysis or gel filtration in most cases. Losses on dialysis can be expected when the molecular weight of the substance is below 15,000, and it is also important to consider the electrical charge of the macromolecules and membranes used in dialysis. When the negatively charged cellophane tubes are used, there is not much loss on dialysis of the positively charged histones (molecular weight about 12,000 to

25,000), whereas the smaller negatively charged proteins escape through cellophane to a considerable degree.

5. Circularly Polarized Light

Circularly polarized light can be produced by resolving a plane-polarized beam into its components (see Chapter II, Sects. 2 and 11). The classical way to accomplish this is to place a thin birefringent crystal in the way of the plane-polarized beam. A birefringent crystal has two axes: one along which the light is retarded relatively little and another along which there is more retardation than along the "fast" axis. A very thin crystalline quartz plate has such properties. Two in-phase vibrations parallel to the axes pass through such crystal with different velocities and on emerging reveal a difference in phase Δ. This phase difference Δ appears to be proportional to the thickness d and to the birefringence $n_s - n_f$, i.e.

$$\Delta = 2 \pi d (n_s - n_f)/\lambda ,$$

where n_s is the refractive index in the most retarded direction, n_f the index in the less retarded direction, and λ the wavelength. A birefringent crystal which introduces a phase difference of $\Delta = \pi/2$ is called a *quarter-wave plate*.

The effect of such a quarter-wave crystal plate on the plane-polarized ray depends on the *angle* between the plane of polarization of the incident ray and the axes of the crystal. If the angle between the plane of polarization and the "fast axis" is 0°, the emerging light will be plane-polarized; and the same will happen when the crystal is rotated through 90° clockwise or counterclockwise (*i.e.* +90° or −90°). However, if the crystal is rotated clockwise through only 45°, the emerging light is right *circularly polarized*. Similarly, if the crystal is rotated through −45° (counterclockwise), the light is left *circularly polarized*. Rotation of the crystal to all intermediate angles will produce elliptically polarized light. In a circularly polarized ray the electric field vector is rotating regularly along the direction of propagation, *i.e.* the vector describes a helix in space.

As in the experiments with plane-polarized light, the presence of circular and elliptical polarization can be tested with appropriate optical devices, *e.g.* a Nicol prism and quarter-wave plate. First, the ray can be tested by the Nicol only; the prism is slowly rotated and the emerging light examined. If its intensity does not change when the Nicol is rotated, the ray is either unpolarized or circularly polarized. (If the light is extinguished, it must be plane-polarized.) A quarter-wave plate inserted ahead of the Nicol prism will convert the circularly polarized ray to plane-polarized, as may be checked with the aid of another Nicol. Elliptically polarized light, in the absence of a quarter-wave plate, produces much the same effect as partially polarized light: upon slow rotation of the Nicol, there is a variation of the intensity of the emerging ray. Again, partial plane polarization can be distinguished from elliptical polarization by interposition of a quarter-wave plate. In the case of elliptical polarization, there will be two angles at which the elliptically polarized ray becomes plane-polarized; these angles are 90° apart and correspond to the coincidence of the "fast axis" of the quarter-wave plate with one or the other axes of the ellipse.

Another classical method of producing circularly polarized light is by total reflection in transparent isotropic materials *(Fresnel rhomb)*. A plane-polarized ray at 45° angle of incidence on the totally reflecting surface of the rhomb is split into two equally intense components with a phase difference Δ (see *e.g.* LOWRY, 1935, p. 237; WOLDBYE, 1967). The phase difference depends on the angle of incidence and index of refraction of the rhomb material. For each additional reflection, the phase difference increases by the same amount, and by proper choice of reflecting material and angle it is possible to construct a device yielding a phase retardation of $\pi/2$, *i.e.* a circularly polarized ray. The direction of rotation of the electric field vector of the circularly polarized ray is reversed by rotating the plane of polarization of the incident ray through 90°. Special glass of high refractive index can be used for studies with visible light and fused silica rhombs of appropriate angles are useful for ultraviolet and visible light. The modern dichrometers and dichrographs, however, employ the much more convenient principle of electrobirefringence, as described in the following section.

6. Circular Dichroism Measurement

The principles of circular dichroism measurement are outlined in the book of VELLUZ, LEGRAND, and GROSJEAN (1965), and in the review articles of WOLDBYE (1967) and BEYCHOK (1967). The measurement of circular dichroism requires monochromatic plane-polarized light, hence some components of a circular dichrograph are the same as in a spectropolarimeter. For this reason, some of the instrument firms which offer spectropolarimeters (Cary, Durrum-Jasco) provide circular dichroism attachments. Instruments for circular dichroism measurement only also are available. These were first provided by the "Société Roussel Jouan", Paris, France; and recently a combination of a Durrum-Jasco dichrograph and spectrophotometer has been offered. Circular dichroism attachments for the Beckman and Cary spectrophotometers also have been offered, *e.g.* by Rehovoth Instruments Ltd., Rehovoth, Israel.

The principal part in the Jouan and Durrum-Jasco dichrographs is a modulator crystal which resolves the plane-polarized monochromatic ray into two circularly polarized components. The modulation is based on the phenomenon that certain uniaxial crystals, such as ammonium dihydrogen phosphate, when subjected to an alternatting current, change their dichroic properties. In a suitable arrangement, the crystal can act as a quarter-wave plate, and in the course of one cycle of alternating voltage the plane-polarized ray is modulated into left, then right, circularly polarized components. The modulator is positioned between polarizer and a sample which absorbs the left and right components to a different degree, and the difference is detected by a sensitive photocell and recorded (Fig. 21). Since the optical density differences are very small, the greatest precision in the electro-optical system is mandatory. Amplification of the weak effects leads to problems of electrical noise, *i.e.* a low signal-to-noise ratio.

The highest sensitivity of the original Jouan Dichrographe was a deflection of 1 cm for an optical density difference of 1.5×10^{-3}, (ΔD, see Sect. 11, Chapter II) and the shortest wavelength which was reached in the ultraviolet was about 210 nm. During the last few years, the instruments have been improved considerably, so that the sensitivity is pushed up to a differential optical density of 1×10^{-5} per 1 cm on the recorder chart. The electrical noise, as in rotation, depends on the absorption of the sample, and the precision depends on the magnitude of the dichroic effect and on the interfering total absorbance. A sample recording obtained with the new Durrum-

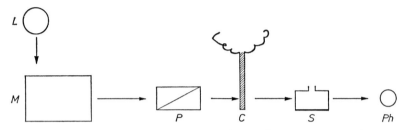

Fig. 21. Schematic sketch of a circular dichrograph. *L* Light source; *M* Monochromator; *P* Polarizer; *C* Electrobirefringent crystal; *S* Sample cell; *Ph* Photocell

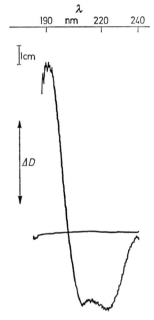

Fig. 22. A circular dichroism recording of the α-helical poly-α-L-glutamic acid in the far ultraviolet. A 0.0091% solution of the polymer in 0.1 *M* sodium fluoride, pH 4.2, was placed in a cell with an optical path of 0.05 cm. The dichroic absorbance scale setting was 1×10^{-4}/cm. (Jirgensons, unpubl.)

Jasco CD-SP Model high-sensitivity dichrograph is shown in Fig. 22. The recording was started in the ultraviolet at 240 nm and continued into the deep ultraviolet. The scale sensitivity was 10^{-4} of ΔD per 1 cm deflection, *e.g.* the total effect at the negative peak at 222 nm is 4×10^{-4} and it is 9×10^{-4} at the positive peak at 191 nm. The positions of the peaks correspond to those wave lengths at which the optical activity changes its sign, *i.e.* the positions where the rotatory dispersion curve crosses the base line (see Sect. 9 in Chapter II). It is now a common practice not to report the direct recording of dichroism but to recalculate the result in terms of ellipticities and plot them on the ordinate in the units of degrees$\times$cm^2/decimole. If the dichroism values are compared, they are usually reduced to a concentration of 1% and an optical path of 1 cm, and denoted by $\Delta D^{1\%}_{1cm}$ or $\Delta E^{1\%}_{1cm}$. Conversion to molar residue ellipticity $[\theta]$ then may be done according to

$$[\theta] = \Delta D^{1\%}_{1cm} \times \bar{M}/10 \times 3300,$$

in which $\bar{M}$ is the mean residual weight of the amino acids in a protein or polyamino acid (see Sect. 11, Chapter II). In our example presented in Fig. 22, the $\Delta D^{1\%}_{1cm}$ value for the positive peak at 191 nm amounted to $19.8 \times 10^3 \times 10^{-4}$, and this, if multiplied by 3300 and the $\bar{M}/10$ value, which in this case is 12.9, yields a molar residue ellipticity of 84,300 (in degrees$\times$cm^2/decimole).

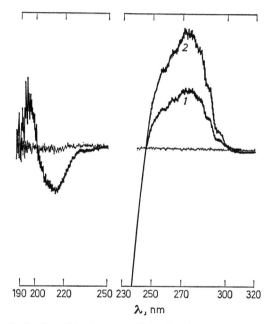

Fig. 23. Tracings of circular dichroism recordings in the 250—320 nm and 190—250 nm spectral zones with the improved Durrum-Jasco dichrograph. As in ORD, the recording is started at the longer wavelength at minimal speed, slit width is monitored, and phototube activating voltage automatically readjusted. The modulator voltage is also adjusted in such a manner that the birefringent crystal acts as a quarter-wave plate at all wavelengths. The specimen in our example is an acid proteolytic enzyme isolated from the mold *Rhizopus chinensis*. The concentration in the 250—320 nm zone was 0.1% (in 0.01 M NaH$_2$PO$_4$, pH 4.4) and the optical path length through the solution was 1.0 cm. Here (on the right-hand side), curve *1* presents the recording at a sensitivity of $1 \cdot 10^{-4}$ differential optical density per 1 cm on the chart, and curve *2* the same at a higher sensitivity of $5 \cdot 10^{-5}$/cm. In the 190—250 nm zone (on the left), the same protein was diluted to 0.01% and the optical path length was reduced to 0.10 cm. The sensitivity in the far ultraviolet was $5 \cdot 10^{-5}$/cm. (JIRGENSONS, unpubl.)

If circular dichroism yields essentially the same structural information as optical rotatory dispersion, why measure circular dichroism? The answer is that circular dichroism measurement resolves the optically active absorption bands more clearly than is possible with rotatory dispersion. In rotatory dispersion the Cotton effects overlap more strongly than in circular dichroism recordings. While in rotatory dispersion measurement the effects are readily measurable far from the absorption bands, which is an advantage, the circular dichroism recording runs on the base line except at the wavelength range of the active band. Moreover, the circular dichroism and ellipticity data can be used for controlling the rotatory dispersion results, especially in the most interesting ultraviolet zone at 190 to 240 nm.

As already indicated, the circular dichroism spectra are much richer in fine structure than the rotatory dispersion curves. This is particularly true in the 250—300 nm zone in which the circular dichroism curves reveal the effects of the various amino acid side chain chromophores. Moreover, improvements in the sensitivity of modern instruments now guarantee a high degree of reproducibility so that it is possible to operate with minute quantities of specimens. Some examples are shown in Fig. 23. Here, in the 250—320 nm spectral zone, curve *1* is a tracing with a sensitivity of $1 \cdot 10^{-4}$ differential optical density $\varDelta D$ per 1 cm on the chart, and curve *2* shows the same amount of material at a higher sensitivity of $5 \cdot 10^{-5} \varDelta D/\text{cm}$. Multiple positive CD peaks are well reproduced with both sensitivities, *i.e.* both recordings show shoulders or peaks at 295, 285—290, 280—283, 277, 271—273, 263—265, and 256—258 nm. Note that in the 190—250 nm region (recording on the left) the amount of the specimen in the beam is 100 times less than at 250—320 nm, *i.e.* the circular dichroism effects in the far-ultraviolet zone are much stronger than those in the near-ultraviolet zone.

Calibration of the dichrographs is done by means of a solution of pure d-10-camphorsulfonic acid in water. This substance has a positive CD band centered at 290.5 nm and a negative band at 191 nm. According to the very careful calibration experiments of CASSIM and YANG (1969), the molar circular dichroism of the 290.5 nm band is $+2.20$ corresponding to a molar ellipticity of $+7260$ degr.cm^2/decimole. Those who are interested in the performance of their instrument in the far ultraviolet are recommended to consider also the negative 191 nm peak. For this band, we have measured a molar ellipticity of $-13,750$ degr. cm^2/decimole of camphorsulfonic acid.

7. Resolution of Circular Dichroism Curves

The positive and negative peaks or bands in the circular dichroism spectra are distribution functions around certain wavelengths; the analysis of a curve consists, first, in the precise determination of these positions. Second, one may be interested in the height of the peak, its width at half-height, and its area. The positions are related to the natural frequencies of the electronic transitions and the heights and areas are related to the rotatory strengths.

While in some spectra the positions and intensities of the peaks are obvious, this is frequently not the case. In Fig. 22, the one positive and the two negative bands are conspicuous; in Fig. 23, however, the situation is much more complex. Resolution of such complex curves into a limited number of certain distribution functions can be accomplished in several ways. The most usual approach is to look for a minimum number of the bell-shaped (normal) Gaussian functions. Digital computers should provide the most accurate answers, but there is no simple program for all cases. Also, one has to realize that the solutions provided by such resolution analyses are not unique, *i.e.* variation of the distribution parameters can produce two or more equally valid answers fitting the experimental curve. To get the most probable answer, the resolution is conducted by chosing certain peak positions, such as the 222 nm negative peak of the helix.

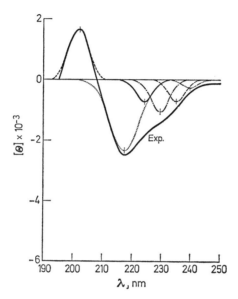

Fig. 24. Resolution of an experimental circular dichroism curve (Exp) into Gaussian bands with the aid of the du Pont 310 Curve Resolver. The dichroism is expressed as mean residual molar ellipticity. (JIRGENSONS, unpubl.)

In recent years, curve resolution has been accomplished chiefly with the aid of analog computers. The *du Pont Model 310 Curve Resolver* is such an analog computer which utilizes a series of function generators, each producing a single distribution function on a cathode-ray tube. The distribution functions, usually as Gaussian peaks, can be positioned, shaped, and summed until a tracing is found that matches the experimental curve. Each of the resolved peaks then can be analyzed separately.

An example of such resolution is shown in Fig. 24. The heavily drawn experimental curve represents circular dichroism of human serum immunoglobulin in the far-ultraviolet spectral zone. The resolution yielded two relatively strong bands, one positive, centered at 202 nm, and on negative, centered at 217 nm; in addition, weak bands were resolved at 224, 229, 235, and 239 nm.

8. Circular Dichroism Measurement in the Vacuum Ultraviolet Region

In the commercial dichrographs which are equipped with quartz optics and flushed with dry nitrogen, the shortest accessible wave length is 182—183 nm. Because of the absorption of the sample, the practical limits are only 185—190 nm. To penetrate the even deeper ultraviolet zone that may hide some interesting but as yet unknown bands, one has to use optical elements that are transparent to such short-wave radiation and extremely thin samples, and remove all gas from the path of the ray. Such devices now have been constructed by several groups of investigators, *e.g.* by W. C. JOHNSON (1971). In his vacuum spectrograph the light source is a hydrogen lamp with a magnesium fluoride window. Dispersion of the ray is effected by a grating monochromator. The monochromatic ray then is plane-polarized by a polarizer made of magnesium fluoride crystal, modulated to its circularly polarized components by a quarter-wave retarder (made of CaF_2), and directed to the cell compartment and photomultiplier. All parts, of course, are evacuated. The sample is usually only one drop of solution sticking between polished CaF_2 windows, the optical paths through the solution being only 0.001—0.01 mm (JOHNSON and TINOCO, 1972). The shortest wavelength reached by this instrument was 135 nm. A shoulder on the big positive CD peak of helical poly-L-glutamic acid was observed at 175—180 nm; and this was the limit for aqueous solution. The α-helical poly-γ-methyl-L-glutamate dissolved in hexafluoro-2-propanol had similar positive bands in the 175—195 nm zone; this system could be examined at even shorter wavelengths, yielding a negative band at about 160 nm followed by a positive peak centered near 140 nm.

Chapter IV

Optical Activity of Amino Acids, Peptides, and Proteins

1. Optical Rotation and Configuration of α-Amino Acids

The optical activity of proteins is due to two factors — the asymmetry of the polypeptide chain and the asymmetry at the α-carbon atoms in the amino acid residues. If the asymmetric order of folding of the chain is destroyed, the protein is still optically active because of the rotatory contributions of all the asymmetric α-carbon atoms in the disorganized chain. Although we are interested chiefly in the effects of the higher-order structures on optical activity, it is also important to know the contributions of the many asymmetric carbon atoms. For this reason, the optical activity of natural α-amino acids will be discussed briefly.

Optical rotation measurements on aqueous solutions of various natural α-amino acids have shown that some of them were levorotatory and some others dextrorotatory. At first, then, it seemed that some might belong to the L and some others to the D series. Even more confusing was the observation that the direction of rotation of a given amino acid depends on the solvent. Extensive studies on the configuration of natural amino acids were conducted in many laboratories and around 1930 it was realized that *all natural amino acids have the same configuration* [1]. This was ascertained by chemical conversions of the amino acids into other substances of known configuration, as well as by biological and physicochemical methods. Important contributions in these endeavors were rendered by measuring the rotatory power of amino acids in solutions containing various amounts of strong acids and alkalies. CLOUGH (1918) was one of the first who observed that acidification of aqueous solutions of several amino acids resulted in a similar shift of rotation, and he indicated that the direction of this shift may be related to configuration. This was investigated systematically on many amino acids by LUTZ and JIRGENSONS (1930, 1931). This work led to a generalization which we originally formulated as follows: "Upon increasing the concentration of an acid (H+) the optical activity of an L-antipode of an α-amino acid becomes more positive, and of the D-antipode more negative" (shift-rule of LUTZ and JIRGENSONS). In other words: *suppression of dissociation of the carboxyl group of an L-α-amino acid results in a positive shift of the optical activity, i.e.*

$$\text{L-acid:} \quad R-\underset{\underset{NH_3^+}{|}}{\overset{\overset{H}{|}}{C}}-COO^- + H^+ \rightleftarrows R-\underset{\underset{NH_3^+}{|}}{\overset{\overset{H}{|}}{C}}-COOH$$

negative or weak   less negative or more
positive rotation   positive rotation

---

[1] Some D-amino acids occur in bacterial cell walls.

This is illustrated in Fig. 25. On the abscissa are plotted the equivalents of hydrochloric acid or sodium hydroxide, on the ordinate the specific rotation at the wavelength of 589 nm. Of the examples presented, leucine, histidine, and tryptophan are

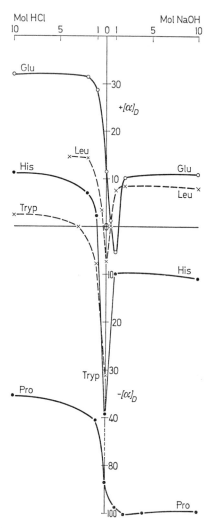

Fig. 25. Dependence of the specific rotation of natural α-amino acids on the number of added moles of HCl or NaOH. Demonstration of the positive shift rule for some L-amino acids. (From Lutz and Jirgensons, 1930, 1931)

levorotatory in an aqueous solution but dextrorotatory in the acidified solutions. The *change* observed on acidification, however, is the same in alle cases. (The same regularity is not observed on adding alkali.) Data on all natural L-amino acids obtained from protein hydrolyzates are presented in Table 3, and expressed in terms of molar rotation. The subject was presented in great detail by GREENSTEIN and WINITZ (1961), who expressed the rotatory shift rule as follows:

$$\text{L-amino acid: } [M]_{\text{acid}} - [M]_{\text{water}} = +\Delta[M]$$
$$\text{D-amino acid: } [M]_{\text{acid}} - [M]_{\text{water}} = -\Delta[M]$$

According to these authors, the rule of the rotational shift holds for the amino acids which have more than one asymmetry center, if the rotatory contributions of the α-carbon are corrected for the contributions of the other asymmetry centers. The rule

Table 3. Molecular rotation, $[M]_D$, of natural L-α-amino acids in aqueous solutions and in 5 N HCl (from GREENSTEIN and WINITZ, 1961)

Amino acid	Mol. wt.	$[M]_D$ in water	$[M]_D$ in 5 N HCl
Alanine	89.1	+ 1.6	+ 13.0
Valine	117.1	+ 6.6	+ 33.1
Leucine	131.1	−14.4	+ 21.0
Isoleucine	131.1	+16.3	+ 51.8
Serine	105.1	− 7.9	+ 15.9
Threonine	119.1	−33.9	− 17.9
Methionine	149.2	−14.9	+ 34.6
Cystine [a]	240.2		−557.4
Cysteine	121.1	−20.0	+ 7.9
Aspartic acid	133.1	+ 6.7	+ 33.8
Glutamic acid	147.1	+17.7	+ 46.8
Ornithine	132.2	+16.0	+ 37.5
Lysine	146.2	+19.7	+ 37.9
Histidine	155.2	−59.8	+ 18.3
Arginine	174.2	+21.8	+ 48.1
Citrulline	175.2	+ 7.0	+ 42.4
β-Phenylalanine	165.1	−57.0	− 7.4
Tyrosine [a]	181.2		− 18.1
Proline	115.1	−99.2	− 69.5
Hydroxyproline	131.1	−99.6	− 66.2
Tryptophan	204.1	−68.8	+ 5.7 (1 N HCl)

[a] Cystine and tyrosine are practically insoluble in water; however it was shown that their optical activity in 5 N HCl was more positive than in diluted HCl.

has been applied extensively in studies on the configuration of many other amino acids and their derivatives. According to KAUZMANN, WALTER, and EYRING (1940), the rule of the positive shift has a sound theoretical foundation.

2. Optical Rotatory Dispersion of α-Amino Acids

The early studies on rotatory dispersion of amino acids in the visible and near ultraviolet are summarized by SCHELLMAN (1960). It appears that L-amino acids with no chromophore or a weak one at the β carbon atom give plain dispersion curves *ascending into positive with decreasing wavelength;* amino acids with a powerful chromophore at the β carbon atom show anomalies in the visible and near ultraviolet. One such exception is *cystine*, which is much more optically active than any other amino acid because of the *asymmetry of the disulfide bond*.

More recently, measurement of the rotatory dispersion of amino acids has been extended into the far ultraviolet. IIZUKA and YANG (1964) measured the rotatory dispersion of 20 L-amino acids at pH 1 between 190 and 600 nm at 27°. All showed a *positive Cotton effect near* 210 nm with a peak at 225 nm and a trough at 193 nm. Cystine exhibited a very deep trough at 206 nm and the whole curve was in the negative part of the chart. The maximum of the positive Cotton effect of cystine was near the base line. Tyrosine and tryptophan had an additional Cotton effect near

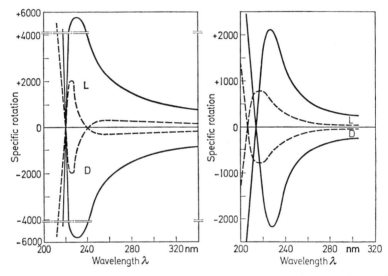

Fig. 26. The rotatory dispersion of phenylalanine (left graph) and alanine (right graph). The continuous curves show the rotatory dispersion of the L- and D-isomers as hydrochlorides, while the dashed curves represent the amino acids in neutral solutions. (From CRAIG and ROY, 1965)

275 nm. The positive Cotton effect is apparently attributable to the carboxyl chromophore which is decisive in the optical activity of these substances. The magnitude of the positive *extremum* near 225 nm, in terms of molar rotation, was between 2000° and 4000° for most of the amino acids, but near zero for proline and cystine.

Rotatory dispersion of six enantiomeric pairs of α-amino acids, both in neutral and in acidic solutions, has been measured by CRAIG and ROY (1965). Their results on alanine and phenylalanine are shown in Fig. 26. The curves of the antipodes are symmetrically opposite, *and the positive shift rule holds for all wavelengths*.

3. Rotatory Contribution of Amino Acid Residues in Peptides

We now return to the problem of the optical activity of a polypeptide devoid of a secondary structure. Is it possible to calculate the optical activity of such a peptide from the known activities of the constituent amino acids? Is the optical activity of an amino acid residue within a polypeptide chain the same as that of the residue at one of the ends of the chain?

BRAND and ERLANGER (1950) compared the optical activity of small L-alanine peptides of known sequence. Some of them were composed of alanine only, while others contained the optically inactive glycine. When the latter was in a terminal position, it was possible to compare the optical activity of the interlocked (endo) alanine residue with the activity of the free alanine or alanine at the N- or C-terminals. BRAND and his colleagues extended this work to the L-lysine peptides. Large activity differences for both alanine and lysine were found depending on their position. This work, chiefly in relation to the optical activity of large polypeptides, has been reviewed by DOTY and GEIDUSCHEK (1953) and by SCHELLMAN and SCHELLMAN (1958). Some of the results of Brand's group are presented in Table 4.

Table 4. Residue rotations of L-alanine at various positions in small peptides. All measurements were made in 0.5 N HCl with sodium light ($\lambda = 589$ nm) (from BRAND et al., see DOTY and GEIDUSCHEK, 1953)

N-terminal position		Endo position	
H . Ala—Gly . COOH	$+29°$	H . Gly .. Ala .. Gly . COOH	$-109°$
H . Ala—Ala . COOH	$+26°$	H . Ala .. Ala .. Ala . COOH	$-130°$
H . Ala—Lys . COOH	$+23°$	H . Gly .. Ala .. Ala . COOH	$-124°$
H . Ala—Lys . Ala . COOH	$+36°$	H . Ala .. Ala . Ala . Ala . COOH	$-169°$

C-terminal position	
H . Gly .. Ala . COOH	$-76°$
H . Lys .. Ala . COOH	$-77°$
H . Gly . Ala . Ala . COOH	$-81°$
H . Ala . Lys . Ala . COOH	$-74°$

Most informative are the data of the first row, *i.e.* on alanine joined only by the optically inactive glycine. At the N-terminal the alanine residue is dextrorotatory, at the C-terminal it is levorotatory, while in the interlocked endo position it is strongly levorotatory. Also, it is obvious that the contribution of the endo residues depends on the substituents on both sides. More recently, SCHECHTER and BERGER (1966) investigated 35 alanine peptides (up to hexapeptides) containing L and D isomers in predetermined positions. It was found that the molar rotation could be expressed to a good approximation as a sum of three different "molar residue rotations": one each for the N-terminal, non-terminal, and C-terminal residue, taking values of opposite sign for the L and D residues. The $[m]_{589}$ values for the non-terminal L-alanine were found to be between -163 and $-178°$.

The effect of amino acid composition on the optical activity of *disordered* proteins was studied recently by TANFORD (1967). He tried to estimate the mean residual specific rotation $[m']_\lambda$ [see Eq. (3), Chapter II] of several proteins from the mean residual rotations of the constituent amino acids. The proteins were dissolved in 6 M

guanidine hydrochloride, and it was ascertained by independent measurements that the macromolecules indeed were fully disorganized. It was assumed that

$$[m']_\lambda = \sum n_i [m_i]_\lambda / \sum n_i \,,$$

where n_i is the number of amino acid residues of type i and $[m_i]_\lambda$ is the non-terminal residual activity contribution. (Since the polypeptide chains in proteins are long, the contributions of the terminal residues could be neglected.) The calculated values were then compared with the observed values at several wavelengths. In spite of the fact that the non-terminal residual rotation values of several amino acids were unknown, the agreement between the calculated and observed $[m']_\lambda$ values for the various proteins was satisfactory. For example, the calculated $[m']_\lambda$ values for lysozyme, serum albumin, and β-lactoglobulin at 300 nm were $-465°$, $-575°$, and $-635°$ respectively and the observed values were $-445°$, $-580°$, and $-645°$. The relatively good agreement between the observed and calculated numbers seems to indicate that the proteins indeed were fully disordered. However, TANFORD was more interested in the different $[m']_\lambda$ values of the various proteins. The results clearly show that the *optical activity of fully disordered proteins is different from protein to protein and that it depends on the amino acid composition.*

This conclusion is of particular importance when one considers the optical activity at the extreme points in the Cotton effect curves of native proteins. Although most of the effect may be caused by the asymmetry of the polypeptide backbone, the intrinsic effect of the many "local" asymmetry centers of the constituent amino acids must also be considered. To get the optical activity contribution of the backbone conformation, the optical activity is corrected for the effect produced by the fully disordered chain. According to TANFORD (1967) and the writer [JIRGENSONS, 1966 (1)], this correction is different for various proteins. According to our measurements, if pure aqueous solutions are examined under the same experimental conditions, the optical activity of phosvitin at the important wave length of 233 nm is $-1,900°$ and that of histone F1 $-3,700°$. Viscosity measurements showed that in pure aqueous solutions both proteins behaved as flexible, random-chain polyelectrolytes. Thus the difference in the optical rotation could be explained only as due to *different amino acid composition.* Phosvitin, a protein of egg yolk, contains serine as one of its major constituents, whereas histone F1 contains unusually large amounts of lysine (27%), alanine (24%), and proline (9.4%). We shall return to these questions in Chapters VII and VIII.

4. Optical Rotation of Proteins at Various pH

ALMQUIST and GREENBERG (1934) were among the first to determine the optical activity of proteins at various pH values. They found that the specific rotation of serum albumin did not change within pH limits of 4 to 11, and that the levorotation increased in strongly acidic and strongly alkaline solutions. Considering the acid solutions, and comparing this with the behavior of natural amino acids (see Sect. 1), it is obvious that the optical activity change in the albumin is not caused by the same molecular events as in amino acids. Later extensive studies in many laboratories showed that the optical activity change in proteins at various pH is caused by *changes in conformation.* The viscosity of the protein solutions increases in strongly acid and alkaline solutions, and this increase is caused by the unfolding or swelling of the globular macromolecules. This unfolding and disorganization of the original secondary and tertiary structure is called *denaturation* (KAUZMANN, 1959; JOLY, 1965). Treating a native globular protein with dilute strong acids or alkalies seldom effects a complete loss of the higher-order structures; a more drastic treatment, *e.g.* with concentrated acids, may cause amino acids and peptides to be split off before denaturation is complete. For this reason, acids and alkalies are not the best denaturing agents. Instead, 8 to 10 M urea or 5 to 6 M guanidine hydrochloride are preferred.

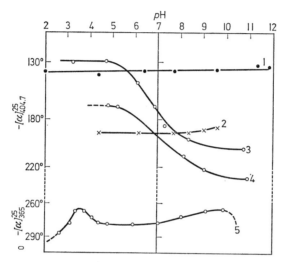

Fig. 27. Dependence of the specific rotation of proteins on pH. Curve *1*, lysozyme; *2* ribonuclease; *3* phosvitin; *4* pepsin; *5* bovine serum albumin. [JIRGENSONS, 1958 (2)]

A comprehensive review on the change of specific rotation of various proteins on denaturation with acids, alkalies, and many other denaturing agents can be found in the excellent book of JOLY (1965), and a recent review was published by TANFORD (1968). The large and complex proteins which have a quaternary structure, such as a serum macroglobulin or glutamic dehydrogenase, are more sensitive to acids and alkalies than are the simple small proteins.

In the early studies on the dependence of optical activity of proteins on pH, low-efficiency sodium light was used. When photoelectric spectropolarimetry became available, more accurate investigation was possible by using ultraviolet light. Fig. 27 shows some data obtained by the writer [JIRGENSONS, 1958 (2)]. Within the pH limits indicated in the illustration, the optical activity of lysozyme (curve *1*) and ribonuclease (curve *2*) was practically independent of pH. However, phosvitin (curve *3*) and pepsin (curve *4*) became more levorotatory when the pH was raised from 5 to 7 and beyond. Both phosvitin and pepsin contain many free acid groups at their side chains; addition of alkali enhances dissociation of these groups, resulting in high charge density, mutual repulsion, and stretching of the polypeptide chains. While lysozyme and ribonuclease are resistant to such change in conformation, a slight but definite structural change could be observed in serum albumin at pH 3.5 to 3.8 and 8 to 10. These latter changes were not detected in earlier work when sodium light was used, but they are well reproducible with ultraviolet light (curve 5). These weak effects have been studied in detail by FOSTER (1960).

It is important to realize that changes in dissociation alone, if they are not followed by changes in conformation, have an insignificant effect on the optical activity of proteins. This is the case with pepsin in acid solutions [JIRGENSONS, 1958 (2)]. Within the pH limits of 1 to 5 the conformation and optical activity do not change, although the degree of ionization is changed.

The effects of acidity on the optical activity and conformation are very pronounced in some polyamino acids, such as poly-α-L-glutamic acid (YANG and DOTY, 1957). This polymer has carboxyl groups at each residue, whereas in proteins the ionogenic groups are only at some of the amino acid residues in the polypeptide chain. The conformation of polyamino acids will be discussed in the next two chapters.

5. Circular Dichroism of Amino Acids

Solutions of simple aliphatic amino acids show CD effects only on irradiation with light of very short wavelength because there is no significant absorption of waves longer than 215 nm. However, the aromatic amino acids and cystine have absorption bands not only in the very deep ultraviolet zone but also near 220 nm and at 250—300 nm (see Sect. 11, Chapter II). The fine structure of both the absorption and CD spectra of these amino acids and their derivatives have been studied recently in some detail. The purpose of these studies is to identify the specific R-groups (side chains) in the CD spectra of proteins and to apply this information to a better understanding of specific structural changes as expressed in the spectra.

Important work on the fine structure of absorption and CD spectra of aromatic amino acids has been done by STRICKLAND and associates (e.g. STRICKLAND, HORWITZ, and BILLUPS, 1969). The absorption and circular dichroism of phenylalanine, tyrosine, and tryptophan, and their derivatives were tested and compared at temperatures as low as the boiling temperature of nitrogen (77° K). To prevent freezing of the solutions, the amino acids were dissolved in water-glycerol and other solvent mixtures and the cells and the instrument were appropriately modified. The CD spectrum of L-tyrosine showed a relatively strong positive peak at 275 nm and another, somewhat weaker band was observed at 281.5 nm. The related absorption maxima were observed at 276.3 nm and 282.6 nm, respectively (HORWITZ, STRICKLAND, and BILLUPS, 1970). The broad CD band diminished at shorter wavelengths and almost vanished at 250 nm. Circular dichroism of L-tryptophan and its derivatives (e.g. N-acetyltryptophan ethyl ester) was measurable at 300 nm and definite positive CD maxima were resolved at 290 nm and 283 nm, with several more peaks at 270—278 nm (STRICKLAND, HORWITZ, and BILLUPS, 1969). Phenylalanine and its derivatives had relatively weak CD bands, e.g. at 261 and 268 nm (STRICKLAND, KAY, and SHANNON, 1970). Cystine has a weak CD band at 278 nm, a band at 255 nm, and a strong negative band at 199 nm; the 255 nm band can be positive or negative, depending on the dihedral angle of the inherently asymmetric disulfide bond (COLEMAN and BLOUT, 1968).

In the above mentioned papers, STRICKLAND and colleagues reported low-temperature CD data also on a few proteins. It was found that the rotatory strength of the CD bands increased on cooling in all cases. However, while the 290 nm band strength of N-acetyl-L-tryptophanamide increased 18 times on cooling from 298° K to 77° K, in chymotrypsinogen the same band showed only a 25% intensification on cooling in the same temperature interval. This indicates that in the case of free amino acids the molecules can assume a large variety of conformational isomers causing overlap and cancellation of the effects, whereas in a protein the rotation of the R-groups is restricted. However, a slight mobility of the R-groups is allowed at room temperature, as in the case of complete rigidity there should be no temperature effect on the CD bands. The allowed mobility of an R-group, of course, depends on the rigidity of the protein macromolecule and its location in the macromolecule. The surface groups have more mobility than those in the interior, and in unfolded (denatured) proteins the side chains have more freedom than within the framework of a globular protein.

Similar considerations appear to be valid for smaller molecules, such as open-chain amino acids and peptides versus cyclic peptides and anhydropeptides or diketopiperazines. STRICKLAND et al. (1970) showed that the increase in CD intensity on cooling a cyclic anhydrodipeptide is much less than for the open chain derivatives of the amino acids. Apparently such cyclic structures have less conformational freedom than linear peptides, as can be demonstrated also by models. It is interesting that in the cyclic anhydrodipeptides that contained aromatic residues the CD bands were located approximately in the same positions as in the open chain molecules but the sign and rotatory strength were different. For example, in a solution of cyclic L-Trp-L-Trp in methanol-glycerol (9:1, v/v), negative CD bands were found at 290.5, 283.5, and 265 nm, the 290.5 nm band being the strongest. Coupling effects of two aromatic rings in the same relatively rigid molecule also were observed by the same authors.

The study of small peptides by chiroptical methods has expanded recently to a considerable degree. The reason for this interest is to understand better the CD spectra of proteins, especially those of low helix content which display unusual CD curves. It is important to identify the conformational changes in the R-group arrangement in various biologically important reactions. Of the many papers which have been published recently on CD of peptides, two more will be mentioned, as they consider the molecules under physiological conditions. SCHECHTER et al. (1971) studied the CD spectra of open-chain oligopeptides composed of L-tyrosine, L-alanine, and L-glutamic acid. The peptides were dissolved in 0.15 M aqueous NaCl containing 0.02 M sodium phosphate, pH 7.4, and the CD spectra were obtained at 27° C. The tripeptide displayed a relatively strong positive band centered at 227 nm and a weak and broad positive peak with a maximum at 270 nm. Moreover, it was found that the CD spectra of the oligopeptides (L-Tyr-L-Ala-L-Glu)$_n$ change in a systematic manner with the increase of the refractive index n (see Chapter VI). ZIEGLER and BUSH (1971) investigated the circular dichroism of cyclic hexapeptides having one or two side chains. The aqueous solutions of these peptides were studied at room temperature. A cyclic peptide composed of 5 glycine (optically inactive) and one L-tyrosine residue had a strong absorbance at 192 nm, a weaker one at 224 nm, and a very weak band at 275 nm (the molar extinction coefficients per average residue molecular weight were 11,900, 1,525, and 220, respectively). This peptide showed a strong negative CD band at 198 nm, positive CD bands at 210 and 228 nm, and a relatively very weak and broad negative band centered at 275 nm. The authors assigned the 198, 228, and 275 nm bands to the neutral tyrosyl side chain of the cyclopeptide. Another cyclic hexapeptide c-Gly$_2$-L-Tyr-Gly$_2$-L-His had a similarly shaped CD curve, only the 198 nm Cotton effect had a higher amplitude than the c-Gly$_5$-L-Tyr peptide and the bands in the 210—240 nm zone were also negative. Regarding the tyrosine CD band centered at 270—275 nm, it is noteworthy that, depending on the vicinal effects of the other constituents, it can be positive or negative. A detailed theoretical study on tyrosine and its isomers was published by Hooker and Schellman (1970).

Chapter V

The Optical Rotatory Dispersion of Polyamino Acids and Proteins. Measurements in the Visible and Near Ultraviolet Spectral Zones

1. Rotatory Dispersion of Polyamino Acids as Protein Models

Synthetic polyamino acids have been used extensively in the attempts to solve the problems of protein conformation. Poly-γ-benzyl-α-L-glutamic acid was used for testing the Moffitt theory as early as 1956 (Sect. 7, Chapter II). Poly-α-L-glutamic acid and poly-α-L-lysine are some of the other useful model polymers.

The polyamino acids are synthesized by various methods, mostly from the N-carboxy-α-amino acid anhydrides (see the review of KATCHALSKI and SELA, 1958). The latter polymerize easily, as follows:

$$n\,\underset{\underset{\displaystyle O=C\!\!-\!\!-\!\!-\!\!O}{|}}{HN\!-\!\overset{\overset{\displaystyle R}{|}}{CH}\!-\!CO} \quad \longrightarrow \quad -\left(-HN\!-\!\overset{\overset{\displaystyle R}{|}}{CH}\!-\!CO\right)_{n}\!-\,+n\,CO_2\,.$$

The N-carboxy-α-amino acid anhydrides themselves are prepared chiefly from the amino acids and phosgene.

All polyamino acids are polymolecular, *i.e.* they are inhomogeneous with respect to molecular weight. The samples can be made more homogeneous by fractionation, but complete homogeneity can not be attained. Depending on the case, the mean molecular weight of the major fraction may be from several thousand to several hundred thousand. The solubility of these polymers varies depending on the amino acid composition and molecular weight. The resemblance to proteins is not limited to structural features but is extended to similarities in hydrolytic decomposition by enzymes (*e.g.* SIMONS, FASMAN, and BLOUT, 1961) and a limited antigenicity. According to GILL and DOTY (1961), a simple polyglutamic acid or polylysine is not antigenic, but copolymers of L-glutamic acid, L-lysine, and L-tyrosine act as antigens in rabbits. A comprehensive review on the synthesis and properties of polyamino acids has been provided by KATCHALSKI et al. (1964).

The optical rotatory properties of some of the most important polyamino acids are listed in Table 5. A large number of these protein models have the Moffitt constants, b_0, within the limits of -580 to -660. This indicates the possible presence of the α-helical conformation. The presence of long and stiff rodlike particles in the solutions of these polymers has indeed been confirmed by other methods, such as viscosity, flow birefringence, light scattering, and electron microscopy. The most important work in this field was performed in Prof. DOTY's laboratory at Harvard University from 1956 to 1961, and the findings have been reviewed extensively, notably by URNES and DOTY (1961).

While a strongly negative b_0 of approximately -600 is indicative of the right-handed α-helix, it is not so easy to interpret those cases where the b_0 is near zero or positive. In the instance of poly-α-L-glutamate in neutral or weakly alkaline aqueous solution a low positive b_0 is known to signify disordered structures. In the cases of a high positive b_0, such as in the last three examples in Table 5, three possibilities have

Table 5. Optical activity, Drude constants (λ_c), and Moffitt constants (a_0 and b_0) of some synthetic polyamino acids. All b_0 values have been obtained with $\lambda_0=212$

Polymer	Solvent	Optical rotation	λ_c	a_0	b_0	References
Poly-γ-benzyl-α-L-glutamate	Ethylene dichloride			$+205$	-635	1
Do.	Dioxane			$+135$	-630	1
Do.	Chloroform			$+250$	-625	1
Do.	Dimethylformamide			$+200$	-660	1
Poly-α-L-glutamic acid	Dimethylformamide			$+135$	-580	1
Do.	Water, pH 4.4				-610	2
Do.	Water, pH 10.5		202		$+ 50$	2
Poly-α-L-lysine	Water, 0.2 M NaBr, pH 11.9			$- 30$	-650	3
Do.	Water, 0.2 M NaBr, pH 6.8		210	-980	0	3
Copoly-L-lysine-L-glutamic acid (equimolar)	2-Chloroethanol	$[\alpha]_D+12$			-636	4
Poly-L-methionine	Chloroform	$[\alpha]_{546}+22$			-630	5
Poly-L-tyrosine	Water, 0.15 M NaCl, pH 10.85	$[\alpha]_D+150$	237		$+540$	6
Poly-L-tryptophan	Dimethylformamide	$[\alpha]_{546}+218$	235		$+410$	7
Poly-β-benzyl-L-aspartate	Chloroform				$+665$	8

Ref. 1, MOFFITT and YANG (1956); Ref. 2, BLOUT (1960); Ref. 3, APPLEQUIST and DOTY (1961); Ref. 4, DOTY, IMAHORI, and KLEMPERER (1958); Ref. 5, FASMAN (1961); Ref. 6, COOMBES, KATSCHALSKY, and DOTY (1960); Ref. 7, SELA, STEINBERG, and DANIEL (1961); Ref. 8, BRADBURY et al. (1960).

been considered: (1) helical conformation of opposite (left-handed) screw sense, (2) ordered conformations other than a α-helix, and (3) dominance of the side chain chromophores. The left-handed helix is believed to be very likely in poly-benzyl-aspartate, whereas in poly-tyrosine and poly-tryptophan the side group chromophores probably overshadow the effects produced by the peptide groups. It is noteworthy that the tyrosine and tryptophan polymers have a strong dextrorotatory power at the long wavelengths, and that their Drude constant (λ_c) is normal (Table 5). Of the other poly-amino acids, the L-proline polymers represent a special group which will be discussed in Chapter IX.

In the course of later studies, *aqueous solutions* of poly-α-L-glutamic acid and its co-polymers, and poly-α-L-lysine received the greatest attention, because they are comparable to proteins with respect to the solvent. The solvent effects are evident from the few cases shown in Table 5. Even in aqueous systems, the b_0 is strongly affected by acidity. However, more detailed studies showed that acidity is not the only factor, and that the rotatory properties, *e.g.* of poly-α-L-glutamic acid, *also depend on* the presence of neutral salts and other neutral additives. FASMAN, LINDBLOW, and BODENHEIMER (1964) found that sodium chloride, lithium bromide, urea, and ethylene glycol strongly affect the b_0 of polyglutamic acid and its co-polymer with L-leucine. For example, upon addition of NaCl (to 0.2 M) to an aqueous solution of polyglutamic acid at pH 4.88, the b_0 was changed from -625 to -460. In the presence of 1 M LiBr, at pH 4.88, the b_0 dropped even to zero. It is also interesting that 8 M urea was unable to disorganize the polymer much, because the b_0 was changed only to -480 (pH 4.88). However, most surprising was the strong negative shift of the b_0 in a mixed solvent containing two parts aqueous 0.2 M NaCl and one part ethylene glycol. A b_0 of -700 to -708 was found for the polyglutamic acid and a b_0 of -860 for the copolymer of glutamic acid and leucine (pH 5.18). If, as is usually assumed, a b_0 of -630 (± 30) means 100% α-helical conformation then a still more negative value is absurd. Such findings have considerably undermined confidence in the usefulness of the Moffitt constants of polyamino acids as standards for the study of protein conformation. It should also be mentioned that the change of pH and ionic strength may result in a different change of b_0 in proteins than in polyglutamate or other polymer. For example, addition of a *salt* (*e.g.* NaCl) to an aqueous solution of histone results in a negative shift of b_0.

To avoid the strong solvent effects, only aqueous solutions of poly-α-L-glutamate (or its acid form) can be considered as standards. A thorough examination of the dependence of the b_0 of these solutions on pH has been done by TOMIMATSU, VITTELLO, and GAFFIELD (1966). In the presence of 0.05 M NaCl, they found a b_0 of -670 at pH 4.2 to 4.5; the b_0 changed to -590 at pH 5.0, and the negative value dropped further to -345 at pH 5.3 (λ_0 in all cases was 212). In the presence of 0.01 M NaCl, the b_0 of the same sample was -640 at a pH of 5.0. These authors found that the polymer aggregated in acid solutions at pH values below 4.5. This aggregation did not affect the b_0, although the other Moffitt constant, a_0, was considerably influenced thereby.

URNES (1963) determined the b_0 values of copolymers of L-glutamic acid (95% glutamic acid + 5% L-tyrosine; and 40% L-glutamic acid, 30% L-alanine, and 30% L-lysine) in different spectral zones. He found that in the long-wave zone (λ 300 to 600 nm) the λ_0 of 212 yielded the best straight lines, whereas extension of the measurements to the shorter wave lengths of 250 to 300 nm required a higher λ_0 value of 216 to 220 in order to have a straight-line relationship. He plotted the $[m']_\lambda (\lambda^2 - \lambda_0^2)/\lambda_0^2$ values against $\lambda_0^2/(\lambda^2 - \lambda_0^2)$. When a λ_0 higher than 212 was used in the graphical estimation of b_0, a less negative value was found for the latter. For example, the magnitude of b_0 for the various polymers was -525 to -592 with a λ_0 of 216, and was -337 to -403 with a λ_0 of 220. All these data refer to weakly acid solutions in which the polymers were believed to be completely α-helical. In neutral or alkaline solutions the polymers are disordered and low positive values of b_0 were found in all cases.

In Table 6 the most probable b_0 values for the 100% helical standards of poly-α-L-glutamic acid and its copolymers are compiled. These are mean values for aqueous solutions of low ionic strength obtained in many laboratories, including our own measurements. The possible variations, the $\pm$ deviations indicated, are smaller if the same sample is compared under identical conditions.

Table 6. Moffitt constants, b_0, for 100% α-helical poly-α-L-glutamic acid

$$\left[\begin{array}{c} -HN-CH-CO- \\ | \\ CH_2 \\ | \\ CH_2 \\ | \\ COOH \end{array} \right]_n$$

determined by using different values for the parameter λ_0

Wavelength range, nm	λ_0, nm	b_0, degrees [a]
300—700	212	−640 (±35)
250—400	216	−560 (±25)
240—350	220	−375 (±15)

[a] Mean values from different sources (see also JIRGENSONS, 1965).

2. Rotatory Dispersion of Proteins

The specific rotations and the dispersion constants of many proteins have been determined. In Table 7 are compiled data obtained from measurements with visible and ultraviolet light. The fibrous structural proteins, the glycoproteins, histones, and the colored proteins are excluded from this compilation. Their rotatory properties will be discussed later. The Drude constants, λ_c, were found as described in Sect. 6 of Chapter II. In applying the Moffitt equation (Sect. 7, Chapter II), some of the b_0 values have been obtained by using $[\alpha]_\lambda$ instead of $[m']_\lambda$ in the expression $[m']_\lambda (\lambda^2 - \lambda_0^2)$ on the ordinate. This method yields a somewhat more negative b_0 than the use of the corrected specific rotation. As the variable on the abscissa is used $1/(\lambda^2 - \lambda_0^2)$ or $\lambda_0^2/(\lambda^2 - \lambda_0^2)$ or $\lambda_0^4/(\lambda^2 - \lambda_0^2)$. For a wide range of wavelengths, by using the λ_0 of 212, the $\lambda^2 - \lambda_0^2$ values have been computed and tabulated by FASMAN (1963). One also finds there the data for $1/(\lambda^2 - \lambda^2{}_{212})$ as well as the magnitudes of $\lambda^2 - \lambda^2{}_{216}$ and $1/(\lambda^2 - \lambda^2{}_{216})$ for some wavelengths in the ultraviolet zone. Some authors use the expression $[m'_\lambda] (\lambda^2 - \lambda_0^2)/\lambda_0$ on the ordinate, and in many papers the symbol $[R']$ or $[R']_\lambda$ is used instead of $[m']_\lambda$. Fig. 28 shows an example of the Moffitt plot with $\lambda_0 = 212$, and the decimal adjustment needed for this way of plotting is indicated [1]. The corresponding wavelengths also are shown on the abscissa. The results represent the behavior of a Bence Jones protein in the native state (line *1*) and in the presence of various denaturing agents (lines *2*, *3*, and *4*). The positive slope of line *1* indicates that this protein is nonhelical. Line *2* shows the rotatory dispersion of the same protein after modification of its conformation by treatment with sodium dodecyl sulfate, and the negative slope indicates the presence of the α-helix. In case *3*, the denaturation was

---

[1] The rearrangement of the Moffitt-Yang equation and sample calculations of the b_0 values are presented by FASMAN (1963). According to the method of plotting shown in Fig. 28, the positive slope of line *1* is 0.67 and the $b_0 = 0.67 \times 100 = 67$, because the abscissa values in this plot are multipled by 10^{-4} and the ordinate values by 10^{-6}. Similarly, the slope of line *2* is −1.23 and the $b_0 = -123$. The b_0 values are always expressed in angular degrees, although this is often not mentioned.

accomplished by using a mixture of dodecyl sulfate and urea, and in case *4* the protein was treated with 8 *M* urea (room temperature, pH 7.3). In this latter case line *4* is horizontal, *i.e.* the Moffitt constant $b_0 = 0$, which indicates complete disorganization (JIRGENSONS, 1963). Since the optical rotation is negative, all of the lines are below the abscissa, and the lower the position of the line the more negative is the rotation. In the Moffitt plots, this position of the lines is characterized by the constant a_0, which can be obtained from the intercept.

The limitations of the Moffitt method in estimating the α-helix content are evident in Fig. 28. The Bence Jones proteins, like many other important proteins, are known

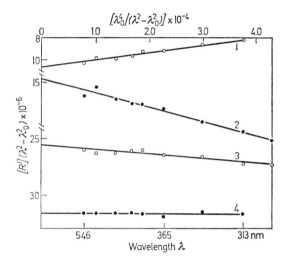

Fig. 28. The Moffitt-Yang plots of native (curve *1*) and modified Bence Jones protein. The Moffitt constant, b_0, is obtained from the slopes, the constant a_0 from the intercepts of the lines. (From JIRGENSONS, 1963)

to be devoid of the α-helix, having other conformations (see (Chapter VIII). The rotatory dispersion data from the visible and near ultraviolet, when analyzed by either the Drude or the Moffitt method, are unable to reveal the β structures. However, in a relatively large number of proteins the α-helix content can be estimated with a fair degree of accuracy from the b_0 values. Thus, this can be done for the proteins which have high negative b_0's (see Table 7). For example, native serum albumin has a b_0 of -280, and its helix content (*H*) can be estimated as

$$H = 100 \frac{280}{640} = 44\%.$$

This is the way the helix content has been calculated by many authors. However, all such estimates must be considered as *crude approximations*, even if the α-helix and disorder were the only factors which determine the b_0 values. In the calculation, it is assumed that the b_0's of fully disordered, flexible, random-chain proteins and fully disordered chains of the standard polyamino acid, are zero. This assumption may not

Table 7. Optical rotation and rotatory dispersion of proteins from measurements in the visible and near ultraviolet spectral zones. The Moffitt parameter λ_0 was 212 nm. The symbol D indicates the sodium line of $\lambda = 589$ nm

Protein	λ nm	$[\alpha]_\lambda$ degr.	λ_c nm	a_0 degr.	b_0 degr.	References
Albumin, bovine serum, in water	D	− 63	264	−287	−288	URNES, DOTY (1961)
Albumin, human serum, pH 5.7	D	− 63	263	−297	−281	CALLAGHAN, MARTIN (1962)
β-Amylase, pH 8.1	D	− 26	277		−195	JIRGENSONS [1961 (1), 1964]
Aspergillopeptidase A, in water	D	− 35	207	−211	0	ICHISHIMA, YOSHIDA (1966)
Bence Jones protein, type K	436	− 87	195	−292	+ 43	HAMAGUCHI, MIGITA (1964)
Bence Jones protein, type L-I	436	− 83	212	−265	0	Do.
Bence Jones protein, type L-II	436	−112	215	−355	− 12	Do.
Carbonic anhydrase B, pH 7.0	D	− 53	204	−359	+ 26	EDSALL et al. (1966)
Do., in 8 M urea	D	− 93	210	−622	+ 5	Do.
Carbonic anhydrase C	D	− 56	209	−374	+ 3	Do.
Do., in 8 M urea	D	− 92	203	−627	+ 60	Do.
Carboxypeptidase A, pH 9.4	D	− 21	269		−190	JIRGENSONS [1961 (1), 1964]
β-Casein A, pH 7.0—7.4	D	−112	218	−705	− 30	HERSKOVITS (1966)
β-Casein B, pH 7.0—7.4	D	−118	215	−735	− 20	Do.
α-Chymotrypsin, pH 7.0	D	− 66	236	−349	− 99	RAVAL, SCHELLMAN (1965)
Chymotrypsinogen A, pH 3.0	D	− 73	231	−436	− 86	SMILLIE, ENENKEL, KAY (1966)
Chymotrypsinogen B, pH 3.0	D	− 78	232	−465	− 97	Do.
Enolase, pH 7.4	D	− 36	277		−172	MALMSTROM (1962), JIRGENSONS (1964)
Fetuin, pH 7.8	D	− 60	228	−365	−125 [a]	VERPOORTE, KAY (1966)
Fibrinogen, pH 6—7	D	− 49	260	−266	−214	MIHALYI (1965)
Do. in 5 M guanidine-HCl	D	− 86	220	−517	− 39	Do.
Fumarase, muscle, pH 7.3	578	− 23		− 81	−311	KANAREK, HILL (1964)
γ-Globulin, bovine serum, pH 7.8	D	− 46	213		− 20	CHOWDHURY, JOHNSON (1963)
γ-Globulin, human serum, pH 6—8	546	− 58	211		0	JIRGENSONS [1958 (1), 1963]
γ-Globulin, rabbit, pH 7.3	D	− 58		−275	+ 35	GOULD, GILL, DOTY (1964)
Do., pH 11.6	D	− 78		−330	0	Do.
Do., pH 1.9	D	− 76		−350	+ 10	Do.
γ-Globulin, rabbit, Fab fragment, pH 7.4	546	− 42			+ 64	STEINER, LOWEY (1966)
γ-Globulin, rabbit, Fc fragment	546	− 86			0	Do.
Glutamic dehydrogenase, pH=7.1	546	+ 13 [b]				JIRGENSONS [1961 (3)]

[a] A λ_0 of 205 was used in this case.
[b] This is an exceptional case of dextrorotation in the visible spectrum.

Table 7 (cont.)

Protein	λ nm	$[\alpha]_\lambda$ degr.	λ_C nm	a_0 degr.	b_0 degr.	References
Glyceraldehyde-3-phos-phate dehydrogenase, pH 7.7	D	− 39	254		−158	HAVSTEEN (1965)
Do., pH 8.5	D	− 36	286	−211	−197	BOLOTINA et al. (1966)
Do., in 6 M urea	D	−131	222	−780	− 46	Do.
Growth hormone, bovine, pH 9.5	D	− 28	277		−263	JIRGENSONS (1960, 1963)
Hexokinase, yeast, pH 7.0	D	− 35	283		−325	KENKARE, COLOWICK (1965)
Insulin, pH 2.3	D	− 30	266	−150	−200	HERSKOVITS, MESCANTI (1965)
Do., pH 8.5	D	− 34		−194	−124	MARKUS (1964)
Do., pH 8.5, 8 M urea	D	−120		−692	0	Do.
α-Lactalbumin, pH 7.0	D	− 75	256	−355	−235	HERSKOVITS, MESCANTI (1965)
Do., in 8 M urea, pH 7.4	D	−111	222	−710	− 10	Do.
Lactic acid dehydro-genase, chicken heart, H_4, pH 6.9	D	− 31	277		−264	DI SABATO, KAPLAN (1965)
Do., 0.035 M Na-dodecyl sulfate	D	− 52	265		−259	Do.
Do., in 6 M urea	D	− 69	230		−162	Do.
Lactic acid dehydro-genase, human heart, I, pH 7.0	D	− 41	266	−236	−193	JAENICKE (1964)
β-Lactoglobulin, pH 5.6	D	− 27		−169	− 66	TANFORD, DE, TAGGART (1960)
Do., in 8 M urea	D	− 99		−663	− 51	Do.
Lipoprotein, human, HDL_2, pH 8.6	D	− 14	289	− 18	−514	SCANU (1965)
Lysozyme (muramidase), H_2O	D	− 44	254	−250	−183	URNES, DOTY (1961)
Do.	D	− 55	242	−295	−150	HERSKOVITS, MESCANTI (1965)
Malic dehydrogenase, pH 7	D	− 34	270	−153	−221	THORNE, KAPLAN (1963)
Macroglobulin, human, sample Ba, pH 8.6	578	− 33	238	−192	0	WETTER, JANKE, HERTENSTEIN (1966)
Do., sample Gr, pH 8.6	578	− 35	184	−204	+ 52	Do.
Ovalbumin, hen, in water	D	− 31	266	−177	−198	URNES, DOTY (1961)
Papain, pH 3.6	D	− 63	231		−100	JIRGENSONS (1963)
Pepsin, in water, pH 5.2	D	− 69	216		0	JIRGENSONS (1958, 1963), PERLMANN (1959)
Pepsinogen, pH 7.7	400	−156	236		−160	PERLMANN (1963)
Phosvitin, egg yolk, pH 6—7	D	− 72	213		0	JIRGENSONS [1958 (1), 1966 (1)]
Ribonuclease, in water	D	− 72	233		−100	JIRGENSONS (1963)
Do., in 8 M urea	546	−122	221		0	Do.
Soybean protein, 11 S, pH 7.6	440	− 78	222	−253	− 28	FUKUSHIMA (1965)
Thyroglobulin, pH 6.0	550	− 62	230		− 60	EDELHOCH, METZGER (1961)

Table 7 (cont.)

Protein	λ nm	$[\alpha]_\lambda$ degr.	λ_c nm	a_0 degr.	b_0 degr.	References
Trypsin, pH 3.0	D	-40	227		-47	Jirgensons [1961 (1), and unpubl.]
Trypsin, inhibitor, soy bean, pH 6—8	D	-79	217		0	Jirgensons [1962 (2)]
Do., in 0.10 M Na-decyl sulfate	546	-71			-126	Do.

be true in all cases, as indicated in the previous chapter. Also, accurate determination of the Moffitt constants on various specimens of poly-α-L-glutamates in neutral and slightly alkaline solutions have yielded *low positive values for the disordered chains.* The b_0 values were indeed close to zero (0 to $+12$) if a λ_0 of 212 was used in the plots, but they were positive (52 to 73) when a higher λ_0 of 216 or 220 was employed. Thus the correction is especially important when the b_0's were determined using values of 216 to 220 for the parameter λ_0.

3. Limitations of the Drude and Moffitt Methods in the Elucidation of Protein Conformation

A detailed examination of the data in Table 7 shows that many proteins, even if in native state, have low magnitudes for the Drude constants (λ_c) and b_0 values close to zero. Aspergillopeptidase, Bence Jones proteins, carbonic anhydrases, the γ-globulins (immunoglobulins), macroglobulins, pepsin, trypsin, and soybean trypsin inhibitor are some of the proteins having such optical rotatory properties Knowing that the b_0 of disordered polyglutamate is close to zero, one might conclude that the macromolecules of the mentioned proteins are disordered flexible chains. This would be a quite incorrect conclusion. It is well known that these proteins are rigid and that at least some of them have ordered secondary and tertiary structures. Thus the limitations of the Drude and Moffitt treatments for these cases are quite apparent.

Some of the proteins which have low negative b_0 values (e.g. chymotrypsin, β-lactoglobulin, and ribonuclease) are known to have the β conformation in their macromolecules. This, however, cannot be reliably estimated from the Moffitt plots, although some authors have attempted it (Troitskii, 1965).

In Table 7, a few data are given on the rotatory properties of denatured proteins. In concentrated urea solutions (6 to 8 M) the $[\alpha]_D$ values become more negative, the Drude constant λ_c decreases, and the Moffitt constant b_0 becomes less negative than the respective values for the native protein. This has been clearly demonstrated, e.g. with insulin, lactalbumin, and ribonuclease. This is interpreted as loss of the ordered secondary and tertiary structures. Tanford (1967) has shown that a complete transition from the rigidly ordered into fully disordered random conformation occurs in 6 M guanidine hydrochloride.

Treatment with detergents affects the rotatory properties of proteins differently than does denaturation with urea or guanidine salts. If the native protein has a high negative b_0, and thus presumably has a high α-helix content, a detergent has only a slight effect on λ_c and b_0. Such is the case with lactic acid dehydrogenase. If, however, the native protein has a b_0 value close to zero, as in the case of soybean trypsin inhibitor, the detergent provokes a negative shift of b_0. This has been interpreted as a transition resulting in α-helix formation [JIRGENSONS, 1961 (2), 1962 (2), 1963]. A complete compilation of rotatory dispersion data of denatured proteins (the older literature up to 1962) can be found in the book of JOLY (1965).

The problems encountered in interpreting the observed changes of the b_0 values on denaturation are not the major difficulties of the Moffitt method. One fundamental difficulty is that *straight lines in the Moffitt plots are often obtained only in a narrow spectral zone*. The b_0 value then depends on the choice of the wavelength limits (see *e.g.* KRONMAN, BLUM, and HOLMES, 1965). Many uncertainties thus are possible, and many false conclusions have been drawn. The reasons for the non-linearity are absorption bands and weak Cotton effects associated with them. The safest wavelength range for the determination of b_0 for colorless proteins is 300 to 700 nm. The drawback here is that accurate results in this spectral zone are obtainable only with large amounts of a protein.

4. Amino Acid Composition and Conformation of Proteins

GOLDSACK (1969) tabulated the Moffitt parameters of 107 proteins and tried to find out whether there is any relationship between these rotatory dispersion constants and amino acid composition of the proteins. A linear relationship indeed was found between b_0 and the sum of residue percentages of alanine, arginine, aspartic acid, glutamic acid, cysteine (and half-cystine), leucine, and lysine. Proteins rich in these amino acids had higher negative b_0 values than proteins containing relatively little of these amino acids. Since a high negative b_0 is indicative of high α-helix content, one may conclude that sequences of alanine, leucine, etc. favor the helix formation. On the contrary, proteins having a high content of the other amino acids (*e.g.* proline and serine) have b_0 values near zero. The methods of amino acid sequence determination have recently been perfected to a high degree (cf. NEEDLEMAN, 1970), and complete sequences are known for many proteins (cf. SOBER, 1970). Although it is theoretically possible to determine conformation from known sequences, *exact* predictions of the various orders have not been very successful. One reason for these difficulties is the aperiodic distribution of the helix formers among the anti-helix residues in the chain. There are problems of geometric fit of the different size side chains and their interactions (see *e.g.* LOW et al., 1968; FINKELSTEIN and PTITSYN, 1971). X-ray crystallography studies have now revealed the complete three-dimensional structures of many proteins (Chapters VII and VIII) so that the predictions can be thoroughly tested. X-ray structural analysis has disclosed an unexpected variety of conformations in globular proteins.

Chapter VI

The Far Ultraviolet Cotton Effects of Synthetic Polyamino Acids

1. The α-Helical and Random Conformations of Poly-α-L-glutamic Acid and Poly-α-L-lysine

The far ultraviolet Cotton effects of synthetic polyamino acids have been studied extensively in several laboratories. Since the conformation of some of the polymers was established by other methods, attempts have been made to correlate the Cotton effect data with conformation. Poly-α-L-glutamic acid and poly-α-L-lysine are the most thoroughly investigated polymers, and they have proved their value as models for the study of protein conformation. Very important in this field are the discoveries of BLOUT, SCHMIER, and SIMMONS (1962), SARKAR and DOTY (1966), and GREENFIELD, DAVIDSON, and FASMAN (1967).

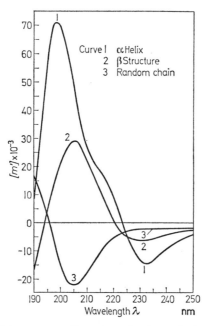

Fig. 29. The far ultraviolet rotatory dispersion of poly-α-L-lysine in α-helical conformation (curve *1*), β conformation (curve *2*), and disordered form (curve *3*). (From GREENFIELD, DAVIDSON, and FASMAN, 1967)

In this section we shall discuss the Cotton effect characteristics of α-*helical* and *disordered* (flexible random chain, also called random coil) poly-α-L-glutamic acid and poly-α-L-lysine. Fig. 29 shows the Cotton effects of poly-α-L-lysine in three different conformations: the right-handed α-helical form, the antiparallel β conformation, and the disordered (random coil) form. The peak and trough values in terms of the corrected mean residual specific rotation, $[m']_\lambda$, for this polymer and for poly-α-L-glutamic acid are given in Table 8. These data are the best mean values calculated from our own measurements and data from other laboratories [SARKAR and DOTY,

Table 8. The positions of the peaks and troughs and the magnitudes of the positive and negative extrema of poly-α-L-glutamic acid and poly-α-L-lysine in the α-helical and random conformations

Polymer	Minimum at nm	$[m']_{min}$ degrees	Maximum at nm	$[m']_{max}$ degrees
Poly-α-L-glutamic acid				
$[-HN-CH(CH_2CH_2COOH)-CO-]_n$				
helical, pH 4.3—4.8	233	−16,200	198—199	+72,000
Do., random, pH 7—8	205	−17,000	191	+23,000
Poly-α-lysine				
$[-HN-CH(CH_2CH_2CH_2CH_2NH_2)-CO-]_n$				
helical, pH 11—11.5	233	−14,600	198—199	+63,000
Do., random, pH 5—7	205	−21,700	190	+18,500

1966; JIRGENSONS, 1966 (1); DAVIDSON and FASMAN, 1967; GREENFIELD, DAVIDSON, and FASMAN, 1967; YANG, 1967 (2)]. The variations, as reported by different workers are quite considerable, *e.g.* the $[m']_{233}$ values for α-helical polyglutamic acid have been found between −14,000 and −18,000°; and the $[m']_{198}$ values for α-helical polysine have been reported to be between +55,000 (SARKAR and DOTY, 1966) and +71,000° (GREENFIELD, DAVIDSON, and FASMAN, 1967). The standard deviations have been found to be approximately 1000 to 3600° at 190 to 210 nm, and 200 to 600° at 220 to 250 nm.

GREENFIELD, DAVIDSON, and FASMAN (1967), with the aid of a computer program, have calculated a series of far-ultraviolet rotatory dispersion curves for certain combinations of 100% α-helical and 100% random chain (disordered) poly-α-L-lysine on the assumption that no other conformations or side chain (R-group) effects are involved (see Fig. 30). It will be shown in the following chapters that the Cotton effects of many proteins indeed resemble those shown in Fig. 30. However, there are also many other proteins which exhibit Cotton effect curves significantly different from those shown in Fig. 30 (see Chapter VIII). Inspection of Fig. 30 shows that upon a decrease of the α-helical form and increase of the disordered form the trough at 233 nm becomes shallower and the peak at 198 to 199 nm is diminished. While the *position* of the trough at 233 nm does not alter, the position of the peak is shifted to shorter wavelengths with an increase of disorder.

SARKAR and DOTY (1966) have studied also the Cotton effects of a copolymer of polylysine which contained 4% of L-tyrosine, and found that this small amount of tyrosine produced Cotton effects not significantly different from those of pure poly-lysine.

The Cotton effect data of the described synthetic polymers can be used for estimation of α-helix content in those proteins in which the α-helix is the only order of the secondary structure. (Also, the proteins should not contain a very large number of aromatic *R*-groups and disulfide bonds.) The *α-helix content* can be estimated either from the magnitude of the negative extremum at 232 to 233 nm or from the peak at 198 to 199 nm by using the values for the presumably 100% α-helical polyamino

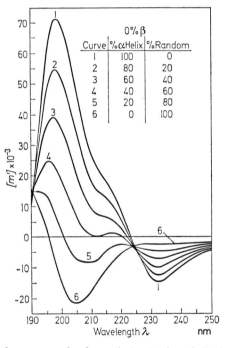

Curve	%αHelix	%Random
1	100	0
2	80	20
3	60	40
4	40	60
5	20	80
6	0	100

0% β

Fig. 30. The Cotton effect curves of poly-α-L-lysine in the α-helical and disordered (random chain) forms, and calculated curves for various proportions of the α-helix and disorder. (From GREENFIELD, DAVIDSON, and FASMAN, 1967)

acids as standards. However, one has also to consider in these estimates the *contribution of the random conformation* [see e.g. YANG, 1967 (1)]. The $[m']_{233}$ values are near $-2000°$ for disordered polyglutamate and near $-2400°$ for polylysine. Thus, at this wave length an average value of $-2200°$ should be used for the correction, *i.e.* the standard would be

$$-16,200 - (-2200) = -14,000.$$

If one uses the polylysine value of $-14,600°$, a lower value is obtained for the standard; however, the helical conformation in polylysine is not as firmly established as in polyglutamate. Therefore, the corrected $-14,000°$ (± 500) value is preferred.

The helix content can be estimated also from the strong positive peak in the ORD curve at 198—199 nm. At this point the rotatory power of 100% helical structure is approximately $+70,000°$. Of course, in this case too we have to consider the contribution of the disordered conformation. For disordered polyglutamate, the $[m']_{199}$ value is near $-3000°$, whereas the corrected specific rotation of polylysine is approximately $-9000°$. For polyglutamic acid the standard value for the peak thus would be

$$+72,000 + 3000 = 75,000.$$

2. ORD of Other Structural Orders

If an aqueous solution of poly-α-L-lysine, pH 11.0 to 11.3, is heated to 50° a transition from the α-helical to the β conformation occurs. The Cotton effect curves thereby change strongly as indicated in Fig. 29. The polymers having the β structures display a trough at 229 to 230 nm and a peak at 204 to 205 nm. The magnitudes of the extrema, as determined on different samples in two laboratories, are compiled in Table 9. The agreement, especially regarding the positions of the trough and peak,

Table 9. The positions and magnitudes of the troughs and peaks of poly-α-L-lysine in the β conformation (aqueous solutions)

Polymer	Trough nm	$[m']_{min}$ degrees	Peak nm	$[m']_{max}$ degrees	References
Poly-α-lysine	230	−6,200	205	+27,000	GREENFIELD, DAVIDSON, FASMAN (1967)
Do.	230	−6,300	204—205	+23,000	SARKAR and DOTY (1966)
Copoly-L-lysine-L-tyrosine (96 : 4)	230	−6,500	205	+21,500	SARKAR and DOTY (1966)

as also the magnitude of rotatory power at the through, is very good. This provides a standard for determining the β conformation in those proteins in which this conformation is the major order. GREENFIELD, DAVIDSON, and FASMAN (1967) have computed curves for various mixtures, i.e. systems composed of certain amounts of the α-helical and β structures, or predominantly random (disordered) polymers containing certain definite proportions of the α- and β-forms. It is noteworthy that some of these curves are similar to the Cotton effect curves of some proteins, as will be discussed in following chapters. However, it must be emphasized that the effects of the α-helical conformation are dominating and the β structures cannot be detected in macromolecules which have only a small amount of the β form. For example, if a polyamino acid or protein contained 80% of the α-helix and 20% of the β form, the Cotton effect curve would not be distinguishable from that produced by 80% helical and 20% disordered chain. Hence these far-ultraviolet rotatory dispersion measurements are useful in detecting the β conformation only in macromolecules in which the helix content is low and the β form is the predominant conformation.

A variety of interesting Cotton effects has been described from observations on poly-amino acid films. BLOUT and SHECHTER (1963) studied axially oriented films of poly-L-iso-leucine and observed a positive extremum at 207 nm. FASMAN and POTTER (1967) compared the far ultraviolet rotatory dispersion of oriented and unoriented films of several polyamino acids. The unoriented films of poly-O-acetyl-L-serine displayed Cotton effects similar to those of the solutions of poly-L-lysine in β form, *i.e.* a trough was observed at 230 nm and a positive peak at 206 nm. However, poly-S-benzyl-L-cysteine, poly-S-methyl-L-cysteine, and poly-L-serine, especially in an unoriented condition, exhibited Cotton effects which had troughs at 233 to 240 nm and peaks at 210 to 215 nm. The curve of the unoriented poly-S-benzyl-L-cysteine had a deep negative trough at 194 nm and the peak was near the base line, whereas the oriented film showed a higher peak. Infrared dichroism and other evidence indicated the presence of the β conformation in all these cases. It seems likely that the peak at 204 to 207 nm might indicate the possible presence of the antiparallel interchain β structure, whereas the peak at 210 to 215 nm would represent the intrachain cross-β form. The problem, how-ever, is far from being solved. The side chain (R-group) effects cannot be disregarded, and the polypeptide backbone conformation has not been sufficiently well ascertained in all cases by other independent methods. Also the theoretical calculations of PYSH (1966) and ROSENHECK and SOMMER (1967) have not been able to give an unequivocal answer as to what sort of β structures produce the observed Cotton effects.

Poly-L-proline is another important polyamino acid which displays Cotton effects dif-fering from those described above. This polymer appears in two forms which were studied by several groups of investigators as models of collagen. According to BLOUT, CARVER, and GROSS (1963), poly-L-proline II shows a strong negative Cotton effect with a negative extremum at 212 to 220 nm and a positive peak at 192 to 195 nm. A film of axially oriented poly-L-proline-I film, however, shows Cotton effect curves approximately opposite to those of poly-L-proline-II in solutions. These Cotton effects are thought to be caused by the poly-L-proline helices and the possibility of opposite (right- and left-handed) helical twists have been considered (BLOUT and SHECHTER, 1963), depending on isomerization and orientation. (The poly-L-proline helix, of course, differs from the α-helix found in poly-L-glutamic acid.) The structure of poly-L-proline, collagen, and gelatin will be discussed in Chapter IX.

3. Circular Dichroism of Polyamino Acids of Various Conformations

The circular dichroism of synthetic polyamino acids has been studied extensively in many laboratories. It has been known for some time that α-helical poly-α-L-glutamic acid exhibits circular dichroism and ellipticity curves having negative CD bands at 205—210 nm and 222 nm and a strong positive band centered at 191 nm (HOLZWARTH and DOTY, 1965). A recording of such a CD curve was shown in Fig. 22. The calculated ellipticities amount to about 84,000 degr. cm²/deci-mole for the 191 nm and 39,000 degr. cm²/decimole for the 222 nm transition. Like the corresponding rotatory dispersion data, these (or similar) ellipticity values have been used as standards for the estimation of the helix content in proteins. For example, HASHIZUME, SHIRAKI, and IMAHORI (1967) measured the CD of several polyamino acids and proteins and estimated the helix content by using the molar residual ellipti-city of 38,000 for the 222 nm band standard. Also, these authors reported the inte-resting observation that the circular dichroism curve of poly-α-D-glutamic acid was a mirror image of the poly-α-L-glutamic acid curve. The far-ultraviolet Cotton effects of the presumably helical poly-L-alanine have been studied by QUADRIFOGLIO and URRY [1968 (1)], who also published CD curves of myoglobin and lysozyme. For the poly-L-alanine, extrema were observed at 221, 207, and 191 nm with ellipticities of

−31,000, −36,000, and +66,000 respectively. The same authors investigated the CD of poly-L-serine [QUADRIFOGLIO and URRY, 1968 (2)]. Circular dichroism of poly-L-lysine has been studied in several laboratories, *e.g.* by GREENFIELD and FASMAN (1969) who also estimated the helix content of proteins from these poly-L-lysine standards. The curves of this polyamino acid in the α-helical, β, and "random" conformations are shown in Fig. 31.

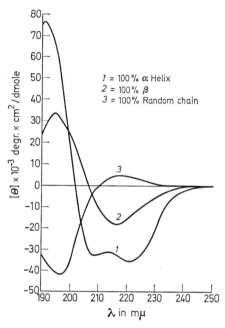

Fig. 31. Circular dichroism spectra of poly-L-lysine in the α-helical, β, and "random" conformation. Taken from GREENFIELD and FASMAN: Biochemistry **9**, 4108 (1969)

More recently, an extensive comparative circular dichroism study on several helical polyamino acids has been reported by CASSIM and YANG (1970). Significant differences in the molar ellipticity peak values (band maxima) were found not only between various presumably α-helical polyamino acids but also between specimens of the same polymer. Some of their data are compiled in Table 10. The peak values of the negative 221—222 nm band vary between −35,800 and −45,900 and those of the positive maximum were between 88,500 and 108,000. Part of the variation may be explained by differences in the refractive index values of the solvents, but the Lorentz factor $3/(n^2+2)$ correction is possible only for solutions in water, because the refractive indices of other solvents in the deep ultraviolet zone are not known. Values for the refractive index, n, increase with decreasing wavelength (Sect. 9 in Chapter II), thus the values of the factor $3/(n^2+2)$ decrease in the deep-ultraviolet zone. For water, the factor is 0.759 at 222 nm and 0.734 at 191 nm. Application of the correction would yield an ellipticity of −31,000 for the 222 nm band maximum of aqueous polyglutamic acid and $+66{,}000$ for the 191 nm band. (Because of the uncertainties involved, CASSIM and YANG (1970) do not recommend applying the Lorentz correction.) There is some uncertainty also about the positions of the peaks. For example, the positions of the 222 nm band maxima in some cases are shifted to 220.7 nm (Table 10). However, when the spectra were resolved into their Gaussian components, the maxima of the 222 nm band in all cases appeared at 223 nm; the position of the negative 208—209 nm band maximum in the resolved curve appeared at 206—207 nm, whereas the position of the positive maximum remained at 190.6—190.8 nm.

Table 10. Circular dichroism peak values of α-helical polyamino acids at 25°.
(After CASSIM and YANG, 1970)

	Residual molar ellipticity $[\Theta]$ at band maximum	Wavelength, nm
Poly-α-L-glutamic acid in water, pH 4.75		
Lot G-44	− 41,900	222.0
	− 39,000	209.1
	+ 88,500	190.6
Lot G-76	− 40,900	221.8
	− 38,900	209.0
	+ 92,000	190.6
Lot G-80	− 39,900	221.8
	− 37,900	208.9
	+ 88,800	190.8
Poly-L-lysine in aqueous 0.5 M NaClO$_4$, pH 6.3	− 35,800	220.7
	− 42,000	208.2
	+ 88,600	190.7
Poly(γ-methyl-L-glutamate) in 2,2,2-trifluoroethanol	− 45,900	220.7
	− 52,700	208.5
	+108,000	190.7

In spite of some uncertainty about the positions and magnitudes of the ellipticity peak values in the α-helical polyamino acids, it is possible to *estimate the helix content of proteins from their CD or ellipticity curves* with a reasonable degree of accuracy if the helix content is high or moderately high (30—80%). The problem of estimating the amount of the pleated-sheet β-structure and other nonhelical conformations in proteins from the polyamino acid standards is much more difficult. This is evident, e.g. from the often quoted publication of GREENFIELD and FASMAN (1969): they computed circular dichroism spectra for various proportions of the α-helical, β, and "random" conformations based on the standards of poly-L-lysine in the three conformations (see Fig. 31). Comparisons on proteins of known conformation (from X-ray structural analysis, see next chapters) showed a reasonably good agreement for proteins of high helix content but poor agreement for the proteins of low helix content. Further analysis of the problem showed that there are two major causes for the discrepancies: first, the unsuitability of heated alkaline poly-L-lysine as a standard for the β conformation, and, second, the fact that in globular proteins the nonrepeating (aperiodic) folds of the polypeptide backbone are not continually changing, flexible random coils. Poly-L-lysine appears in the β form also in neutral 1% sodium dodecyl sulfate solution (LI and SPECTOR, 1969), but the residual molar ellipticity at 216 nm in this solvent is only $-9,600$ degr. cm^2/decimole. According to QUADRIFOGLIO and URRY [1968 (2)], poly-L-serine possesses the pleated sheet conformation in water as well as in several organic solvents; however, the ellipticity curves have a negative peak maximum at 222 nm, while the positive band is centered at 197 nm. The residue molar ellipticity values for the aqueous solutions were $+24,000$ for the 197 nm band and $-5,000$ for the 222 nm band. According to the theoretical work of WOODY (1969), the CD curves of the β-structures in polymers should depend on the length and width of the pleated sheet and whether the arrangement is parallel or antiparallel. Thus the application of the polyamino acid CD standards for the determination of the pleated-sheet conformation in proteins is less reliable than in the case of the α-helical standards. The common features of all the experimental CD curves of the β-structured polymers are: only one negative band of the $n-\pi^*$ transition at 215—222 nm (magnitude approximately $-5,000$ to $-10,000$) and a stronger positive band at 195—200 nm (magnitude about $+30,000$).

There is a similar uncertainty about the polyamino acid standard for the "disordered" or "random-coil" conformation. According to TIFFANY and KRIMM (1969) neither ionized poly-L-glutamate (pH 6—8) nor acidic poly-L-lysine can be considered as flexible random coils but rather as relatively stiff, extended, partially ordered macromolecules. This contention has been confirmed recently by a theoretical analysis of HILTNER, HOPFINGER, and WALTON (1972). The elimination of the electrical charge effects, however, converts the polymers to flexible random coils (TIFFANY and KRIMM, 1969). The CD curves of these truly disordered chains differ strongly from those of the charged polyamino acids (curve 3, Fig. 31). Addition of salts to the aqueous solutions of the supposedly disordered neutral poly-L-glutamate or poly-L-lysine results in disappearance of the weak positive peak at 215—220 nm and flattening of the negative trough at 195—200 nm. The presence of 4—5 M LiClO$_4$, or 4 M CaCl$_2$ is needed to produce the conversion.

According to ROSENKRANZ and SCHOLTAN (1971), the most suitable standard for the pleated-sheet structure is poly-L-lysine in 1% sodium dodecyl sulfate, and for the disordered conformation poly-L-serine in 8 M LiCl. The latter has a very weak dichroism at 210—230 nm and a flat trough centered near 200 nm (magnitude approximately $-7,000$). These standards, however, have been tested for very few proteins.

The other conformations, the rotatory dispersion of which has been briefly described in Sect. 3 of this chapter, have been also investigated in some detail by CD. Of particular interest in this group is poly-L-proline II and the heteropolymers containing also glycine and alanine. According to BROWN, CARVER, and BLOUT (1969), the CD curve of a poly-glycyl-L-prolyl-L-alanine resembles that of collagen in that (Gly-Pro-Ala)$_n$ exhibited CD curves with a weak positive peak at about 220 nm and a strong negative trough near 200 nm (magnitude $-30,000$ to $-44,000$, depending on solvent and temperature).

A summary of the CD spectra of polyamino acids of various conformations is given in Table 11.

Table 11. Positions of circular dichroism band maxima and magnitudes of the residue molar ellipticities of polyamino acids of different conformations

	Position, nm	$[\Theta]$, degr. cm^2/decimole
α-Helix [a]	222	$-39,000$
	208	$-37,000$
	191	$+88,000$
β Pleated-sheet structures	215—222	$-5,000$ to $-10,000$
	195—200	$+24,000$ to $+32,000$
Charged and extended "random" chain	238	-150
	217	$+4,600$
	197	$-42,000$
Random coil, flexible, uncharged	195—210	$-4,000$ to $-12,000$
Poly-L-proline II	226	$+3,000$
	206	$-59,000$

[a] Recently, much higher ellipticity values for 100% α-helical copolymers of L-leucine were reported by CHOU, WELLS, and FASMAN (1972). The highest value for the negative band at 222 nm was $-51,190$ (± 1420) (in 90% methanol at 0°).

A related question is: how long must a polypeptide chain be in order to be able to assume some ordered conformation? SCHECHTER et al. (1971) studied a series of peptides possessing the structure (L-Tyr-L-Ala-L-Glu)$_n$, with $n = 1, 2, 3, 4, 7, 9$, and 13 and compared the CD spectra of these peptides with the spectrum of a high molecular heteropolymer of the same composition. They concluded that in aqueous solutions under physiological conditions, only the polypeptide (L-Tyr-L-Ala-L-Glu)$_{13}$ was partially ordered.

Chapter VII

The Cotton Effects and Conformation of Proteins

1. Attempts to Classify Proteins According to Their Conformation

Several decades ago, STAUDINGER (1935, 1950) suggested that all organic colloids could be classified into spherocolloids and linear colloids. This classification was also applied to proteins, and, according to their macromolecular shape or gross conformation, they were divided into *globular proteins* (spheroproteins) and *fibrous* or *linear proteins* [ASTBURY, 1935; HAUROWITZ, 1936; JIRGENSONS, 1962 (3)]. This classification was very helpful in solving many confusing problems in the colloid chemistry of proteins. In the course of time, much has been learned not only about gross conformation, but also about conformational details within the macromolecules, especially by X-ray diffraction and chiroptical methods. The available facts now make it possible to attempt a more detailed classification with respect to the secondary and tertiary structure or "short range" conformation. The purpose of this section is to propose a tentative classification of globular proteins according to their conformation.

The globular proteins have been previously classified according to conformation into three groups [JIRGENSONS, 1958 (2), 1961 (1); URNES and DOTY, 1961; SCHELLMAN and SCHELLMAN, 1961]: those with a high content of α-helical forms, with a low content of the helix, and those with conformational orders other than the α-helix. The division was made on the basis of optical rotatory dispersion data, mostly from the Drude constants, λ_c. The lines of demarcation used by various authors were somewhat different, although the λ_c value of approximately 210 was accepted as the dividing value between the group of the nonhelical and partially helical proteins. There are no globular proteins which are 90 to 100% α-helical. Application of the semiempirical equation of MOFFITT and YANG (1956), and use of the well-investigated synthetic polyamino acids as standards facilitated the estimation of the α-helix content. According to this approach, most of the globular proteins seemed to have a helix content of 10 to 70%. However, neither the Drude nor the Moffitt-Yang method could be used for the conformational studies of the nonhelical proteins.

New vistas were opened by the discovery of the far ultraviolet Cotton effects of polyamino acids and proteins (SIMMONS and BLOUT, 1960; SIMMONS et al., 1961; BLOUT, SCHMIER, and SIMMONS, 1962). The Cotton effect data for many proteins are now available. They have an advantage over the Drude and Moffitt-Yang constants in that they give some insight into the conformation of the nonhelical proteins and into the macromolecules which have a low content of the α-helix. On the basis of the earlier Cotton effect studies in 1962 to 1965, the writer published a classification of globular proteins [JIRGENSONS, 1966 (3)] in which three subgroups were envisioned in the group of the nonhelical proteins.

Since then several important developments have occurred in the field of protein conformation. They are: (1) X-ray diffraction results on monocrystals of proteins showing conformations other than the α-helix and disorder, (2) discoveries of the β structures in aqueous solutions of several polyamino acids (see Chapter VI) and globular proteins, (3) characterization of fully disordered proteins, and (4) the findings that rigid short range (*e.g.* cyclic) structures can produce Cotton effects somewhat similar to those displayed by the α-helical macromolecules. These developments have led to the reexamination of the earlier views on the estimation of the α-helix content from rotatory dispersion data. Furthermore, the new discoveries call for a revision of the classifications according to conformation. As will be described in detail in Chapter VIII, the nonhelical proteins (Class III) display a great variety of Cotton effects. In several cases the particular peaks and troughs in the curves could be explained as being caused by the β conformations. However, the interpretation of the data was complicated by the discoveries in other laboratories that Cotton effects were produced by aromatic R-groups and disulfide bonds, as well as such rigid short-range structures as cyclic peptides (RUTTENBERG, KING, and CRAIG, 1965) and diketopiperazines (BALASUBRAMANIAN and WETLAUFER, 1966), which cannot form the α-helix for steric reasons. Thus, great caution must be observed in interpreting the data, especially when using proteins of unusual amino acid composition, *e.g.* with a high content of the aromatic amino acids or cystine.

In spite of these difficulties, there is now sufficient material to make it possible for us to envision a tentative new classification of proteins on the basis of conformation. There are many reasons, also theoretical (SCHELLMAN and SCHELLMAN, 1964; SCHERAGA et al., 1967; VENKATACHALAM and RAMACHANDRAN, 1967), which compel one to conclude that the α-helix and aperiodic order are the most important conformations and that diketopiperazine rings and similar constrained structures are absent in ordinary proteins. However, the β structures are becoming more and more conspicuous, and they should be included in the new classification. The new classification is illustrated in Fig. 32.

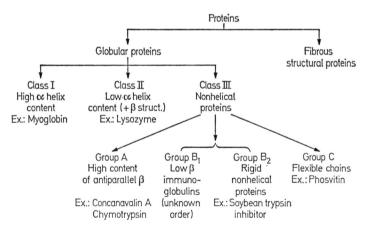

Fig. 32. Classification of proteins according to conformation

2. Comments on Nomenclature

"The customary differentiation of helical and 'random-coil' portions of a native protein is a misapplication of the latter term which can be quite misleading. From what has been said above it will be apparent that a globular conformation is highly specific and certainly not random. Such conformations are compact, whereas the random coil may occupy a domain many times the molar volume" (FLORY, 1967). The solutions of macromolecules which are true random coils or flexible open chains are of high viscosity, and the viscosity increases with increasing molecular weight, whereas the solutions of rigid globular macromolecules are of low viscosity and the viscosity is practically independent of the size of the compact entities. While in the random coil there is a true disorder in that the various segments of the chain assume different positions relative to one another at different moments, the three-dimensional structure in the backbone portions of globular proteins which are neither α-helix nor pleated sheet is *unique* for each protein. During the last decade this has been demonstrated by X-ray structural analysis in many proteins (see Sect. 4 and the next chapter).

How should then the irregular yet *rigid* polypeptide backbone folds and loops be designated? Many expressions have been used, *e.g.* the term "unordered" conformation, "disordered" parts, or "amorphous" segments, or "irregular", as in the previous sentence. The writer suggests that the terms "disordered" and "unordered" should not be used for these conformations, because these structures are specific and unique in each case and thus ordered. The term *aperiodic* or *nonperiodic* conformation seems to be the most appropriate. While we distinguish these structures from the random coil, there are differences in opinion regarding the degree of flexibility of the various structural segments in the macromolecules of globular proteins, such as enzymes (see *e.g.* STRAUB, 1968). Also, some irregular or aperiodic parts may be more flexible than others, *e.g.* the so called hinge regions in the macromolecules of immunoglobulins are relatively flexible, whereas some other chain backbone folds, usually in the interior of the macromolecule, are relatively very rigid.

3. Cotton Effects of Proteins with a High α-Helix Content

Serum albumin, hemoglobin, and myoglobin are some examples of globular proteins which are known to have a high α-helix content. BLOUT, SCHMIER, and SIMMONS (1962) were the first to publish correct data on the far-ultraviolet Cotton effects of serum albumin. The curve appears to be of the same general shape as that of the α-helical poly-L-glutamic acid or helical poly-L-lysine, *i.e.* a trough is observed at 232—233 nm, a shoulder in the positive part of the chart at 215—220 nm, and a high positive peak at 198 nm (curve *1*, Fig. 29).

The ORD data for the globular proteins which display curves with a minimum at 232 to 233 nm and a maximum at 198 to 199 nm are complied in Table 12. The Moffitt constants, b_0, also are included for comparison. The latter, in many cases, have been determined from measurements in the near ultraviolet by using higher values of 216 or 220 for the parameter λ_0. In some cases this was necessitated by the fact that there was not enough material for measurements in the visible spectrum. In some other examples the measurement in the ultraviolet was preferred because of color, *i.e.* absorption bands in the visible spectrum. By working in the ultraviolet, the best straight lines in the Moffitt plots were obtained from data of the 240 to 270 nm or 250 to 350 nm zones, depending on the protein.

The data of Table 12 require a few comments. The Cotton effect curves of many of the proteins mentioned do not have the trough exactly at 233 nm or the peak exactly at 199 nm. For example, *carboxypeptidase-A* has the trough between 233 and 235 nm. *Glutamic dehydrogenase* displays the maximum not at 199 but at 202 to 203 nm. Significant differences in the magnitudes of the positive and negative extrema have been observed depending on the specimen of glutamic dehydrogenase, even by testing preparations of highest purity (JIRGENSONS, 1965). Although contamination with either optically active or inactive extraneous material cannot be strictly excluded, some of the specimens might be denatured. Glutamic dehydrogenase and pancreatic α-amylase have been found to be sensitive to denaturation. In some other cases, such as *insulin*, the rather poor reproducibility may be caused by various degrees of *aggregation*. The peak of the insulin curve is spread from 198 to 202 or even 204 nm. This, as in the case of glutamic dehydrogenase, is suggestive of the presence of the β conformation. The circular dichroism results published by MERCOLA et al. (1967) support the contention that the amount of the α-helix in insulin is moderately low and that β structures may be present. There has been considerable disagreement on the b_0 values of insulin reported from various laboratories (DOTY, 1957; as quoted by URNES and DOTY, 1961; MARKUS, 1964; HERSKOVITS and MESCANTI, 1965), and this disagreement perhaps can be explained in part by differences in aggregation. Moreover, the Moffitt plots are not linear if data are used from the 250 to 300 nm wavelength zone (JIRGENSONS, unpublished results). The β form is known to be present in lysozyme from X-ray diffraction studies. It is noteworthy that a peak or shoulder does not appear in the lysozyme curves if the solutions are tested at pH 5 to 5.5; however, in strongly acid solutions of pH 2 to 2.7 the peak was spread over the 198 to 203 nm region. Similar observations were made on the acid solutions of ovalbumin. The presence of the β structure, beside the helix, is likely in ovalbumin, and the same may be true for some of the other cases, especially those where the peak value is not as high as could be expected from the strongly negative b_0 or the magnitude of the negative extremum (trough).

According to the data of Table 12, denaturation with acid is less effective than denaturation with concentrated solutions of urea. *Human growth hormone* is not disorganized even at pH 1.3 (BEWLEY and LI, 1967). The stability to denaturing agents, of course, varies from protein to protein. However, 8 *M* urea does not disorganize the growth hormone completely (BEWLEY and LI, 1967), nor fibrinogen (BLOUT, 1964), nor the protein component of lipoprotein HDL_2 (SCANU, 1965). According to TANFORD, KAWAHARA, and LAPANJE (1967), proteins are disorganized completely by 6 *M* guanidine hydrochloride at room temperature.

Many authors have attempted to estimate the *helix content in proteins* from either the Moffitt constants, b_0, or the modified Drude equation of SHECHTER and BLOUT (1964), or from the trough or peak values of the Cotton effects. The first two methods have been successful for those proteins which have the α-helix and disorder (random coil) as the only significant conformations. TROITSKII (1965) applied a modified Moffitt equation by assuming a third (presumably β) conformation, but the results are

Table 12. The far ultraviolet rotatory dispersion of globular proteins with a significant amount of the α-helical conformation (Classes I and II). Minima and maxima of the Cotton effect curves, and the Moffitt constants, b_0

Protein	b_0 deg.	λ_0 nm	$[m']_{233}$ deg.	$[m']_{199}$ deg.	References
Adenosine deaminase	−164	216	−7200	+23,000	JIRGENSONS [1966 (3)]
Albumin, bov. serum, H_2O			−9000	+50,000	BLOUT, SCHMIER SIMMONS (1962)
Do.	−210	220	−9000	+35,000	JIRGENSONS (1965)
Albumin, hum. serum, pH 5.1	−286	216	−9100	+37,000	JIRGENSONS [1966 (3)]
α-Amylase, bacterial, pH 10.4	−160	216	−4400	+18,000	JIRGENSONS [1966 (3)]
α-Amylase, pancreat., pH 8—9	−106	220	−4500	+25,000	JIRGENSONS (1965)
Aspartate transcarbamylase,					DRATZ, CALVIN (1966)
pH 6.1			−5200	+23,000	
Carboxypeptidase-A, pH 9.8	−183	216	−5200	+19,500	JIRGENSONS (unpubl.)
Catalase, H_2O	−320	212	−9000	+58,000	YANG, SAMEJIMA (1963)
Do., acid denatured	−115	212	−4900	+36,000	Do.
Do., urea denatured	+ 40	212	−2000		Do.
Conalbumin, pH 5.8	−117	212	−2600	+ 9,200	TOMIMATSU, GAFFIELD (1965)
Creatine kinase, pH 6—7	− 95	220	−5000	+18,000	JIRGENSONS (1965)
Enolase, pH 6—7	−106	220	−5100	+24,000	JIRGENSONS (1965)
Ferrihemoglobin, pH 6	−510	212	−9400	+46,000	BLOUT (1964), BEYCHOK (1964)
Fibrinogen, pH 8.6	−204	212	−5600		BLOUT (1964)
Do., in 8 M urea	−110	212	−3300		Do.
Fumarase, pH 7—7.5	−202	220	−7700	+31,000	JIRGENSONS (1965)
Globin, human	−302	212	−8000		HRKAL, VODRAZKA
Glucose-6-phosphate					JIRGENSONS [1966 (3)]
dehydrogenase	− 84	216	−4600	+13,000	
Glutamic dehydrogenase, pH 7.1	−216	216	−6100	+23,000°	JIRGENSONS [1965, 1966 (2)]
Glutathione reductase, pH 6	−145	220	−5700	+26,000	JIRGENSONS (1965)
Glycerokinase, pH 8	− 99	220	−5300	+17,000	Do.
α-Glycerophosphate dehydro-					Do.
genase, pH 6	−140	220	−6400	+26,000	
Glyoxylate reductase	− 95	220	−4400	+15,000	Do.
Growth hormone, bovine, pH 9.0	−256	216	−7400	+32,300	JIRGENSONS [1966 (2)]
Growth hormone, human,					BEWLEY, LI (1967)
pH 10.1	−295	212	−8000		
Do., pH 7.4	−300	212	−7400		Do.
Do., pH 1.3	−363	212	−8100		Do.
Do., 8 M urea, pH 1.3	−152	212	−3400		Do.
Hemoglobin (see ferrihemoglobin)					
Hexokinase, pH 7	−294	216	−7700	+32,000	JIRGENSONS (1965)
20-β-Hydroxysteroid					JIRGENSONS [1966 (3)]
dehydrogenase, pH 7	−202	216	−6000	+24,000	
Insulin, pH 2.9—3.2	(−265)	212	−5100	+18,000	JIRGENSONS (unpubl.)
Isocitric dehydrogenase	−179	216	−7100	+24,000	JIRGENSONS [1966 (3)]
Lactic acid dehydrogenase,					JIRGENSONS (1965)
pH 6.5—7.5	−165	220	−6300	+32,000	
Lipoprotein, HDL_2, protein					SCANU 1965)
component, pH 8.6	−514	212	−9000	+56,000	
Do., in 8 M urea	−187	212	−3500		Do.

Table 12 (cont.)

Protein	b_0 deg.	λ_0 nm	$[m']_{233}$ deg.	$[m']_{199}$ deg.	References
Lysozyme (muramidase),					TOMIMATSU, GAFFIELD
pH 5.0—5.5	−157	212	−4500	+15,900	(1965)
Do., pH 2.0—2.7			−4700	(+14,700)	JIRGENSONS (unpubl.)
Malate dehydrogenase,					JIRGENSONS (unpubl.)
pig heart muscle, pH 7.0	−185	216	−5300	+22,000	
Myoglobin, whale, H_2O	−292	216	−9300	+42,000	JIRGENSONS [1966 (3)]
Do., pH 7.0	−258	220	−9950		URNES (1963)
Do., H_2O	−250	220	−9000	+47,000	HARRISON, BLOUT
					(1965)
Myoglobin (globin)	−199	220	−7300	+32,000	Do.
Do., in 8 M urea	+143	212	−2500		Do.
Myokinase, rabbit muscle	−305	216	−8100	+29,000	JIRGENSONS (unpubl.)
Nuclease, staphylococcal, pH 7	−258	212	−5600		HEINS et al. (1967)
Ovalbumin, hen, pH 4.8	−146	212	−4000	+16,200	TOMIMATSU,
					GAFFIELD 1965)
Do., pH 2.5	−152	216	−4600	(+22,700)	JIRGENSONS [1966 (2)]
Do., pH 2.5+0.01 M Na-					Do.
dodecyl sulfate	−193	216	−6100	+27,000	
Peroxidase, pH 4.5—5.5	−142	216	−5500	+20,000	Do.
Phosphoglucose isomerase, pH 6	−137	220	−5300	+25,000	JIRGENSONS (1965)
3-Phosphoglycerate kinase,					Do.
pH 7.0	−150	220	−5200	+27,000	
Phosphoglycerate mutase, pH 6	−171	216	−5000	+18,000	JIRGENSONS [1966 (3)]
Phosphorylase (α-glycogen					Do.
phosphorylase), pH 7.9	−139	216	−4600	+20,000	
Phosphotransacetylase, pH 7—8	−206	216	−7200	+29,000	Do.
Pyruvate kinase, pH 6	−120	220	−4700	+21,000	JIRGENSONS (1965)
Toxin, diphtheria, pH 7	− 90	216	−4500	+20,600	Ross (1967)
Triose phosphate isomerase, pH 6	−215	216	−6200	+22,000	JIRGENSONS [1966 (3)]
Xanthine oxidase, pH 7.5—8	−123	216	−3500	+11,000	Do.

not quite convincing. Questions have been raised not only in respect to the *standard values* for 100% α-helical, 100% β, and 100% disordered macromolecules but also about the *corrections to be used for fully disordered proteins of various composition.* The same applies for estimation of the α-helix content from the observed $[m']_{233}$ values of proteins. According to TANFORD et al. (1967), the trough values for fully disordered proteins vary between $-2100°$ (for insulin) to about $-2500°$ for serum albumin and chymotrypsinogen. Even larger differences for the rotatory power of fully disordered chains at 233 nm were observed by the writer [JIRGENSONS, 1966 (1)], *i.e.* $-1900°$ for phosvitin and $-3700°$ for the histone F1. In order to calculate the α-helix content, *e.g.* in bovine serum albumin, one must first to correct the $-9000°$ value for the random chain contribution of $-2500°$ and then divide the result by the standard value of 14,000 for 100% helical chains (Chapter VI). This would yield 6500/14,000 = 0.46 or 46% α-helix.

RIDDIFORD (1966) suggested the use of paramyosin as a standard for estimation of the α-helix content in other proteins, for the following reasons: (1) paramyosin has a rather "normal" amino acid composition, and (2) there is evidence that paramyosin is 100% α-helical. The $[m']_{232}$ value of paramyosin was $-15,400°$ in the native state and $-2100°$ in fully denatured state (7 M guanidine-HCl). Thus the corrected reference value would be $-15,400 - (-2100) = 13,300$. The helix content of serum albumin, if calculated by using this standard, would be 49%.

It is easier to work at 220 to 240 nm than at 190 to 220 nm. For this reason, there are more estimates of the helix content from the $[m']_{233}$ than from the $[m']_{199}$ values. According to RIDDISORD (1966), the peak value of the presumably 100% helical paramyosin is 70,200°. By using this standard, and assuming that the $[m']_{199}$ of the fully disordered paramyosin is near zero, one obtains 35,000/70,200 or approximately 50% helix in the albumin. Thus the agreement in the helix content found in different ways is good.

At this point, it is important to compare the curves calculated for various amounts of helix and disorder (Chapter VI) with the data of Table 12. If only the polypeptide backbone conformation were involved, and if the proteins had the α-helix as the only order, the position of the peak at 199 nm should be shifted to shorter wavelengths when the helix content decreases. In other words, the peaks of the proteins which have a low helix content (low negative b_0, a shallow trough at 233 nm, and a low peak) should be shifted toward the shorter wavelength end. In fact, the opposite is true in many cases. In glutamic dehydrogenase, insulin, lysozyme, and ovalbumin, this opposite shift is quite clear and is indicative of the β structure. In many other cases, even if the peak is low, it remains at 198 to 200 nm. This may be caused either by the amino acid R-group effects or by the presence of small amounts of the β form or both. The weaker the Cotton effects the less reliable the estimation of the helix content. In spite of these uncertainties, the study of the Cotton effects of the largely nonhelical proteins is much more promising for the elucidation of their conformation than the application of the Moffitt method.

4. Rotatory Dispersion, Circular Dichroism, X-Ray Diffraction, and Conformation of Myoglobin, Lysozyme, and Other Proteins

The X-ray diffraction method has been applied to structural studies of proteins with great success. The complete structure of *myoglobin* was disclosed by KENDREW and his coworkers from 1958 to 1962 (KENDREW et al., 1960; KENDREW, 1963). A few years later, the detailed structure of lysozyme molecule was reported by a group of workers headed by PHILLIPS (BLAKE et al., 1965; PHILLIPS, 1966). At the time of this writing, the high resolution X-ray diffraction analysis of several other crystalline proteins (*e.g.* hemoglobin, chymotrypsin, carboxypeptidase, ribonuclease) is either completed or near completion. This makes it possible to compare the conclusions made on the basis of optical rotatory dispersion data with the more straightforward evidence presented by X-ray diffraction.

According to earlier X-ray diffraction data, crystallized myoglobin has a very high α-helix content. The electron density maps of KENDREW et al. (1960) indicated that 75 to 77% of the polypeptide chains in the macromolecules of myoglobin are in the α-helical conformation. Meantime, the rotatory dispersion of the solutions of myoglobin was studied by several groups of investigators, notably URNES (1963). The

Moffitt constants, b_0, were determined from measurements in the ultraviolet and the magnitude of the negative extremum was estimated at 233 nm. It was very encouraging, and even somewhat surprising, that this rotatory dispersion work on myoglobin solutions yielded 75 to 77% α-helix content, *i.e.* exactly as found by the X-ray diffraction method in wet crystals. Two important conclusions were drawn from this comparison: (1) that the conformation is not changed by dissolving the crystal, and (2) that optical rotatory dispersion is a reliable method in the study of secondary and tertiary structure of proteins. During 1963 to 1967, the subject was reinvestigated in several laboratories by means of improved instruments, including refinements in high-resolution X-ray diffraction. These recent studies showed that the amount of perfect α-helical strands in both crystallized and dissolved myoglobin molecules is significantly lower than was previously estimated. A closer look at the electron density maps has revealed that some of the helical strands are more extended or otherwise distorted than they should be in an ideal or perfect right-handed α-helix (ASKONAS et al., 1965; NEMETHY et al., 1967). Thus, it is difficult to present an accurate value for the "perfect helix" even from the X-ray diffraction data. A value of 60 to 65% is now considered more realistic for myoglobin than the earlier higher value.

The Cotton effects of myoglobin have been studied in several laboratories, notably by HARRISON and BLOUT (1965) and BRESLOW et al. (1965). The rotatory dispersion curves in the far ultraviolet are typical for polypeptides having a high content of the right handed α-helix, as shown in Fig. 33. The negative Cotton effect centered at 225 nm has a trough at 233 nm and there is a shoulder at 215 to 220 nm. The strong positive Cotton effect diplays a high positive extremum at 199 nm. If heme is removed from myoglobin, the shape of the curve is not altered but the magnitudes of the extrema are decreased. The curves of myoglobin (sperm whale metmyoglobin) and apomyoglobin (globin, *i.e.* myoglobin without heme) reported by BRESLOW et al. (1965) were very similar to those presented in Fig. 33 (HARRISON and BLOUT, 1965) of similar specimens. Also, there was a close agreement between the $[m']_{233}$ values reported by the two groups of investigators. According to BRESLOW et al., this trough value of myoglobin in various nondenaturing solvents (pH 6.1 to 9.4) was between -8630 and $-8730°$, while HARRISON and BLOUT reported $-9000°$ (± 300). Practically the same value of $-9300°$ was found by the writer [JIRGENSONS, 1966 (2)], while URNES (1963) reported $-9950°$. The small differences are due to a variety of causes, such as errors in optical activity measurement, errors and different methods in concentration determination, and some slight differences in the specimens *(e.g.* the presence of extraneous material). BRESLOW et al. estimated from the $[m']_{233}$ value of myoglobin an α-helix content of 56%, but no correction was made for the rotatory power of fully disordered myoglobin at 233 nm. According to HARRISON and BLOUT, apomyoglobin in 8 M urea is fully disordered, and the $[m']_{233}$ value of the random coil in this case is $-2500°$. (This value may differ somewhat in the presence of heme.) Thus only an approximate estimate of the helix content is possible from the trough values. The mean trough value for native myoglobin, obtained by averaging the data of various authors, is -9100; correction for the disordered chain yields -6600. If one then uses the $14,000°$ value as standard, a helix content of 47% is found. A somewhat higher helix content will be obtained by using a less negative correction for the disordered chain *(e.g.* $-2000°$ instead of $-2500°$) and by using a less negative value for the 100% α-helical polypeptide. According to RIDDIFORD (1966), the $[m']_{233}$

value for paramyosin, which seems to be 100% α-helical, is −15,400°, and the correction for the trough value of the disordered paramyosin is −2100°. Thus, the corrected trough value for this presumably 100% α-helical standard is 13,300°. If one starts now with the same average value of −9100 for the native myoglobin, corrects

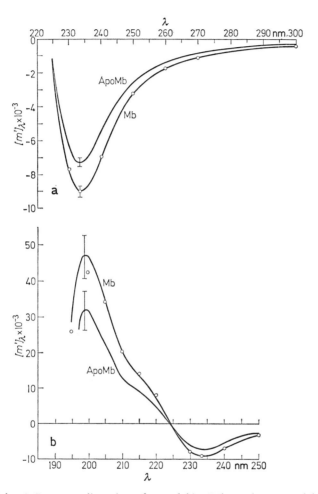

Fig. 33 a and b. a) Rotatory dispersion of myoglobin (Mb) and apomyoglobin (ApoMb). b) Cotton effect curves of myoglobin (Mb) and apomyoglobin (ApoMb) in the far ultraviolet. (From Harrison and Blout, 1965)

it by subtracing 2000, and uses the 13,300 value as standard, the helix content H (in per cent) is

$$H = 100 \ (9100 - 2000)/13,300 = 53\%.$$

Thus, the α-helix content in dissolved native myoglobin, as estimated from the trough values of the negative Cotton effect, is between 47 and 53%. Urnes (1963) discussed another correction in these estimates based on the weak Cotton effects in the visible spectrum; however, this correction does not yield a significantly higher helix content.

The helix content also can be estimated, as indicated in the previous chapter, from the peak values of the positive Cotton effect. The peak values of native myoglobin, according to BRESLOW et al. (1965), were between 35,000 and 38,000° (± 1100), and from these data the helix content was estimated in the range of 56 to 61%. A higher value for the $[m']_{199}$ of 47,000° was reported by HARRISON and BLOUT (1965), but these authors mentioned an uncertainty of $\pm 6000°$. A $[m']_{199}$ of 42,000 (± 1500) was found in our laboratory for the aqueous solutions of native myoglobin. The mean value for the positive extremum thus is near 40,500°. This yields a helix content of 54%, if it is assumed that the rotatory power of disordered myoglobin at 199 nm is near zero and if 75,000 is used as the standard value for 100% α-helical polypeptide chain (see Chapter VI). If the standard value is lowered, to 70,000, as suggested by RIDDIFORD (1966), a higher estimate of 58% is obtained.

In conclusion: according to the Cotton effect data, 47 to 58% of the polypeptide chain in the molecule of dissolved myoglobin is in the α-helical conformation. This is in fair agreement with the most recent X-ray diffraction data, which show 60 to 65% of perfect α-helical strands in the crystals of this protein. Moreover, this analysis shows that only a slight change, if any, occurred upon dissolution of the crystals, and that the *rotatory dispersion can provide reasonably accurate information on conformation.*

The enzyme *lysozyme (muramidase)* is another example where the structural analysis results of X-ray diffraction can be compared with far-ultraviolet rotatory dispersion. The macromolecule of lysozyme, like that of myoglobin, is composed of a single polypeptide chain, and the interpretation of the electron density maps was facilitated by the chemical information on the amino acid sequence (primary structure). Chemically, lysozyme is even simpler than myoglobin, as the latter contains the nonpeptide heme, whereas the former is composed of amino acid residues only. However, the secondary and tertiary structure of lysozyme is more complex than that of myoglobin. A schematic drawing of the polypeptide chain backbone is shown in Fig. 34 (BLAKE et al., 1965; PHILLIPS, 1966). Only a few α-helical regions can be discerned, and some of the helical turns are distorted; thus only an approximate estimate of the α-helical conformation is possible from the X-ray diffraction. There is approximately 30 to 35% of helix, and about 10 to 15% of the extended β structure; the rest of the chain being of irregular or aperiodic conformation. The rectangular blocks indicate the four disulfide bridges in various parts of the chain.

The rotatory dispersion of solutions of lysozyme has been studied in many laboratories, at first by the Moffitt method, and the early work is reviewed briefly by URNES and DOTY (1961) and JIRGENSONS (1963). The b_0 values of lysozyme, computed by using a λ_0 of 212 nm, were in the range -155 to $-180°$, and it was concluded that lysozyme has a moderate α-helix content. More recently, the far-ultraviolet Cotton effects were studied in several laboratories, notably by TOMIMATSU and GAFFIELD (1965). They reported a trough value $[m']_{233}$ of $-4460°$ and a peak value $[m']_{199}$ of $+15,900°$. According to TANFORD et al. (1967), the rotatory power of completely disordered (random coil) lysozyme at 233 nm is $-2300°$. The helix content then should be 4460 minus 2300 divided by 14,000, or 15.5%. A higher value of 21%, however, is obtained from the positive maximum, by using the formerly suggested value of 75,000 as standard and neglecting the correction for the disordered protein at 199 nm. If the value for the standard is decreased from 75,000 to 70,000, the helix content is raised to 23%.

GREENFIELD, DAVIDSON, and FASMAN (1967) have computed the Cotton effect curves of lysozyme by assuming various amounts of the α-helical, β, and random conformations. They found that a curve calculated for 35% α-helix, 10% β structure,

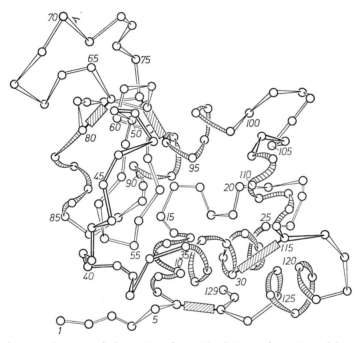

Fig. 34. Schematic drawing of the main polypeptide chain conformation of hen egg-white lysozyme (muramidase) on the basis of X-ray diffraction studies of BLAKE et al. (1965). The rectangular blocks indicate the disulfide bonds. The helical portions are striped

and 55% random chain differs from the really observed curve more than a curve calculated for only 20% α-helix, 32% β, and 48% disordered chain. The R-group effects of the constituent amino acids have been disregarded in these considerations. The disulfide bridges and the high content of tyrosine and tryptophan are some of the remarkable features of lysozyme. Although the far-ultraviolet rotatory dispersion in this case too is determined chiefly by the polypeptide backbone conformation, the amino acid side chains (R-groups), have a stronger effect in lysozyme than in myoglobin.

The Cotton effect curves of lysozyme in nearly neutral or weak acid solutions have the same general shape as the curves of myoglobin. The positive extremum of the β structure, which might be expected to appear at 205 nm, is not observed, because of the overwhelming effect of the helical structures. However, in strong acid solutions (pH 2.5 to 2.7) the peak was spread as a plateau from 198 to 203 nm, and the $[m']$ value of the plateau was 14,700° [JIRGENSONS, 1966 (2)].

Recent measurement of circular dichroism in the far ultraviolet have fully supported the conclusions made on the basis of optical rotatory dispersion data. Fig. 35
shows the ellipticity curves of aqueous solutions of myoglobin and lysozyme obtained
with the sensitive Durrum-Jasco Model CD-SP dichrograph. Both proteins have the
positive peak at 190 to 192 nm which is indicative of the α-helix, and the myoglobin
peak is higher than the lysozyme peak. Moreover, the myoglobin curve has a clearly
formed negative peak at 222 nm, which is another characteristic of the α-helical conformation. The lysozyme curve, instead of having a definite extremum at this wavelength, shows a sloping plateau at 215 to 225 nm. This is compatible with a low helix

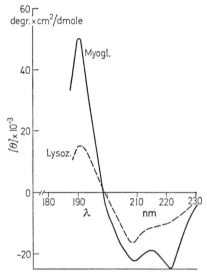

Fig. 35. Ellipticity curves of myoglobin and lysozyme. The negative peaks at 222 nm and the
positive peaks at 190 to 192 nm are characteristic of the α-helix. (JIRGENSONS, unpubl.)

content and with the presence of some β structure. The lysozyme data are in good
agreement with those published by TIMASHEFF et al. (1967). The first far-ultraviolet
ellipticity studies on myoglobin were published by HOLZWARTH and DOTY (1965).
Strong evidence was provided in this work that the ellipticity peaks at 190, 206 to
210, and 222 nm have their origin in the peptide bond conformation. Furthermore,
these authors showed that the fully aperiodic polypeptide chain displayed a negative
circular dichroism centered near 200 nm. Thus the magnitude of the negative trough
at 205 to 210 nm of myoglobin and lysozyme (Fig. 35) probably depends both on the
206 nm transition of the helix and on the effect of the nonperiodic portions of the
polypeptide chain, or perhaps on the effects of distorted helices.

In Table 12 are listed several other proteins which have been sufficiently well
investigated by X-ray diffraction in solid state, so that it is possible to compare the
rotatory dispersion results with the X-ray data. *Ferrihemoglobin* is a prime example.
It is very similar to myoglobin, as now established by X-ray diffraction. It is a protein of high α-helix content. There are only a few other fields of research that have

developed recently to the same extent as the X-ray structural analysis of protein crystals, and there are many reviews on this subject. Some written by experts are those of NORTH and PHILLIPS (1969), DICKERSON and GEIS (1969), and BLOW and STEITZ (1970). Many recent articles on various proteins are compiled in Volume 36 of the Cold Spring Harbor Symposia on Quantitative Biology published in 1972, this issue being entitled "Structure and Function of Proteins at the Three-Dimensional Level". At the time of writing, the conformation of about 20 proteins has been solved to a high degree of precision and the three-dimensional structure of 30 more proteins is known to some degree. A significant amount of the α-helical order has been revealed in about half of all the crystallized proteins studied by this method. *Carboxypeptidase A* has been studied by high resolution (2 A) X-ray structural analysis by LIPSCOMB and associates at HARVARD (see *e.g.* LUDWIG et al., 1967; LIPSCOMB et al., 1969; HARTSUCK and LIPSCOMB, 1971). Approximately 20% α-helix and 25% of the parallel and antiparallel pleated-sheet conformation was revealed in this enzyme. The bacterial protease *subtilisin,* according to WRIGHT et al. (1969) and KRAUT (1971) has about 30% of the α-helix and 20% pleated sheet. The basic *pancreatic trypsin inhibitor,* according to HUBER et al. (1970) has about 25% of the helix and 20% β-form. In the polypeptide backbone of *insulin* BLUNDELL et al. (1971) discovered some 30% of the helix and 20% pleated sheet. *Lactate dehydrogenase* (ADAMS et al., 1970; ROSSMANN et al., 1972) has about 45% of the α-helix and 20% pleated sheet. *Papain* (DRENTH et al., 1968; 1971) has a lesser helix content of 25% and about 15% of the β conformation. In all cases the ordered periodic stretches are short and the helices are distorted to some degree, and the rest of the backbone is in aperiodic folds and loops unique for each protein. Another example, the macromolecular model of the polypeptide backbone of papain, is shown in Fig. 36.

Since the rotatory dispersion and circular dichroism of these proteins has also been investigated, it is now possible to compare and evaluate the potentialities and precision of the chiroptic methods even better than before. In Table 13 are compiled data on the CD (ellipticity) bands which are characteristic of the presence of the α-helical order in proteins. As we see, the helical order is indicated, *e.g.* in insulin, carboxypeptidase, lactate dehydrogenase, and papain. Also, the CD data indicate more helix in myoglobin than in lactate dehydrogenase, and even less of it in papain, all of which is in accord with the X-ray structural analysis. Fig. 37 shows the far-ultraviolet CD spectrum of papain together with the spectra of two other proteins (JIRGENSONS, 1970). The mean residue molar ellipticity of papain is about $-10,000$ at 222 nm, which indicates about 20—25% α-helix in the protein. The protein represented by curve 2, a bacterial amylase, is apparently of lower helix content than papain; and the protein of curve 3 (deoxyribonuclease) appears to be of very low helix content, the negative band at 215—218 nm and positive peak near 195 nm being indicative of the pleated-sheet conformation.

As X-ray structural analysis has now disclosed the conformation of many proteins, we can more readily understand why the polyamino acids do not serve quite well as protein models. Aside from the R-group effects and inadequacy of the "random-coil" model for the aperiodic folds, we see that *the helical segments are short and distorted.* STRAUSS, GORDON, and WALLACH (1969) point out this in a CD study on myoglobin, hemoglobin and lysozyme; they also emphasize the fact that many peptide chromophores in a globular protein are buried in the apolar, highly polarizable core of the

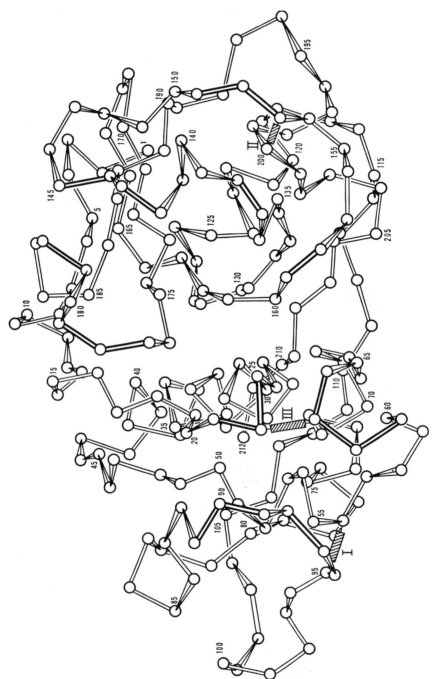

Fig. 36. The polypeptide chain backbone of papain. The circles represent the α-carbon atoms in each of the 212 residues in the macro-molecule. The striped rectangles show the positions of the disulfide bonds. (Courtesy of D. J. Drenth)

macromolecule. A similar conclusion was reached also by SINGHAL and ATASSI (1970); they compared the rotatory dispersion and CD of the various fragments of myoglobin with the spectra of the whole myoglobin. The helix content in the separated fragments was found to be much less than X-ray diffraction analysis indicated in the same sequences in the whole myoglobin. Thus long-range interactions of the various parts in the macromolecule play a major role in its stabilization.

The circular dichroism spectra in the far-ultraviolet spectral zone of 185—250 nm express not only the peptide backbone effects but also the effects caused by the aromatic side chain chromophores and disulfide bonds. For example, the shoulder in the papain curve at 200 nm (curve *1*, Fig. 37) is probably due to these side chain

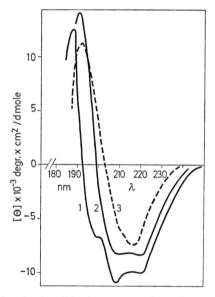

Fig. 37. The far ultraviolet circular dichroism spectra of papain (curve *1*), bacterial amylase (curve *2*), and deoxyribonuclease (curve *3*). The CD effects are expressed in terms of mean residual molar ellipticities. (From JIRGENSONS, 1970)

effects, although the aperiodic folds may also be contributory. Cotton effects in the 220—230 nm region can be caused by the tyrosine chromophores. These *R*-group effects have been little investigated. In many cases they are not observed at all, because of the overwhelming effects of the helices and because of the relatively small number of these chromophores in comparison to the number of peptide bonds. Moreover, the *R*-group effects can be of opposite sign and cancel out.

Considering the structural complexities in proteins, it is no surprise that some of the predictions of conformation based on chiroptic data were not confirmed by X-ray structural analysis. Ferricytochrome c is one such case. According to the data of Table 13, one would predict about 25 to 30% of α-helix, whereas only about 15% has been found by high resolution X-ray diffraction (DICKERSON et al., 1971). In the same article, DICKERSON et al. discuss in some detail the possibilities of predicting the

Table 13. Far-ultraviolet circular dichroism spectra of globular proteins with a significant amount of α-helical conformation

Protein	pH	Position of band max., nm	$[\Theta]$ deg. cm²/ decimole	Reference
Albumin, bovine serum	5.8	222	−15,800	TIMASHEFF et al. (1967)
		210	−17,200	
		191	+34,700	
Albumin, human serum	5.1	222	−18,200	JIRGENSONS (1970)
		209	−16,500	
		191	+39,500	
Aldolase	8.5	222	−14,300	FASMAN, HOVING, and
		210	−13,800	TIMASHEFF (1970)
Alkaline phosphatase	7.0	222	− 6,600	APPLEBURY and
		209	− 6,600	COLEMAN (1969)
		195	+ 9,700	
D-Amino acid oxidase	8.3	222	−12,100	AKI et al. (1966)
α-Amylase, bacterial	5.9	223	− 8,300	JIRGENSONS (1970)
		210	− 8,000	
		192—195	+14,400	
Arginine kinase	7.7	220	−12,000	ORIOL and LANDON
		207	−15,000	(1970)
L-Asparaginase, Escher. coli	7.3	222	− 8,500	FRANK et al. (1970)
		208	− 8,000	
		192—193	+10,500	
Aspartate aminotransferase	7.4	222	−12,000	BERTLAND and KAPLAN
		209	−13,000	(1970)
Carboxypeptidase A	7.2	222	− 7,000	JIRGENSONS (unpubl.)
		210	− 9,000	
		191—195	+14,000	
Citrate synthase	8.2	222	−20,000	WU and YANG (1970)
		210	−21,000	
		191	+42,000	
Creatine kinase	7.7	220	−11,400	ORIOL and LANDON
		207	−14,300	(1970)
Cytochrome b₂, yeast	7.0	220	−10,000	TSONG and
		208	−11,000	STURTEVANT (1969)
		191	+15,000	
Cytochrome c, ferri	7.0	222	−11,200	MYER (1968)
		210	− 9,400	
		190—195	+11,000	
Enolase	6.0	221	−15,700	JIRGENSONS (1970)
		209	−15,700	
		191	+25,500	
Epimerase in 0.02 M Tris		222	−13,000	BERTLAND and
		210	−11,000	KALCKAR (1968)
Fibrinogen	7.4	222	−10,000	BUDZYNSKI (1971)
		209	−10,500	
		190	+23,000	
Flavodoxin, Clostridial	7.0	222	−12,000	EDMONDSON and
		210	−10,000	TOLLIN (1971)
Fructose diphosphatase	6.0	222	−14,000	TAMBURRO et al. (1970)
		209	−14,000	

Table 13 (cont.)

Protein	pH	Position of band max., nm	$[\Theta]$ deg. cm²/ decimole	Reference
Fumarase	7.3	222	−18,700	SORENSEN and
		209	−18,800	HERSKOVITS (1972)
		193	+45,500	
Gliadin, wheat	3.1	222	− 8,300	CLUSKEY and WU
		206	−14,000	(1971)
		190	+11,900	
Growth hormone, bovine	9.5	220—225	−20,000	SONENBERG and
		208	−22,000	BEYCHOK (1971)
		190—194	+35,000	
Growth hormone, human	8.2	221	−19,900	BEWLEY and LI (1971)
		209	−22,400	
Hemerythrin, *Sipunculus*	7.0	220	−21,500	BOSSA et al. (1970)
		210	−22,250	
		195	+58,200	
Hemoglobin, human	7.0	222	−23,000	SUGITA et al. (1968)
		209	−22,000	
Hemoglobin, lamprey	7.0	222	−16,000	SUGITA et al. (1968)
		209	−16,500	
Hexokinase	6.4	221	−14,400	JIRGENSONS (1970)
		209	−13,200	
		192	+25,400	
Histidine ammonia lyase	7.3	222	−13,500	KLEE (1970)
		209	−14,000	
Insulin	6.8	222	−10,100	MENENDEZ and
		209	−12,300	HERSKOVITS (1970)
Isocitrate dehydrogenase	7.0	222	−10,100	CHUNG and FRANSEN
		209	− 8,400	(1969)
		193	+17,400	
Lactalbumin	7.0	220	−11,800	KRONMAN (1968)
		208	−12,600	
		191	+24,200	
Lactate dehydrogenase, muscle, band I	5.9	222	−15,900	JIRGENSONS (unpubl.)
		209	−15,100	
		191	+26,500	
Lipase, *Aspergillus*	6.1	220	− 9,400	JIRGENSONS (1970)
		210	− 8,700	
		192	+17,000	
Lipoamide dehydrogenase	7.6	222	−14,000	BRADY and BEYCHOK
		208	−18,000	(1969)
		193	+18,000	
Lipoprotein, β, high density	6—8	222	−21,000	SCANU and HIRZ
		208	−20,000	(1968)
		193	+47,000	
Lipoprotein, low density	8,6	219	−13,000	SCANU et al. (1969)
		206	−10,000	
Lysozyme (muramidase)	5.7	222	− 7,400	TIMASHEFF et al. (1967)
		209	−10,600	
		193	+12,900	
Myoglobin, whale	6.6	222	−23,500	JIRGENSONS (1970)
		209	−22,000	
		191	+52,000	

Table 13 (cont.)

Protein	pH	Position of band max., nm	[Θ] deg. cm²/ decimole	Reference
Ovalbumin, hen egg	7.5	222	−10,000	GORBUNOFF (1969)
		210	− 9,000	
		192	+20,000	
Papain	3.9	222	−10,000	JIRGENSONS (1970)
		209	−11,200	
		200	− 7,000	
		189	+12,900	
Peroxidase, radish	10.6	222	−13,000	HAMAGUCHI et al.
		209	−14,000	(1969)
Phospholipase A2, pancreatic	7.4	222	−15,000	SCANU et al. (1969)
		209	−14,000	
		192	+25,000	
Phospholipase A, snake venom	7.0	224	−15,000	KAWAUCHI et al.
		210	−13,000	(1971)
Polymerase, RNA-polym., Escherichia	7.5	220	−12,800	NOVAK and DOTY
		208	−13,600	(1970)
Rhodopsin	6.5—7	221	−19,000	SHICHI et al. (1969)
		210	−17,800	
Somatomammotropin, human	8.2	220	−16,700	BEWLEY and LI (1971)
		209	−18,300	
Transferrin	8.0	221	− 6,800	NAGY and LEHRER
		210	− 7,200	(1972)
Tyrosinase, mushroom	7.0	220	− 8,000	DUCKWORTH and
		206	−14,000	COLEMAN (1970)

three-dimensional structure from amino acid sequences and show that these predictions are only partially successful.

It must be emphasized that only *high-resolution* X-ray structural analysis is able to distinguish a helical strand from pleated sheet and pinpoint the location of the *R*-groups. Low-resolution work often shows dense segments that may or may not be helices. Thus there appeared to be a high helix content in carbonic anhydrase examined at 5—6 A resolution, and there was apparent disagreement between chiroptical data and X-ray crystallography. However, further high-resolution structural analysis revealed a very low helix content (KANNAN et al., 1972), and this was in agreement with earlier rotatory dispersion and CD data.

Whether there is any difference in conformation as between a protein in crystal or in solution is a question which has been discussed and reviewed by RUPLEY (1969). On the basis of experimental evidence, he concludes that globular proteins do not change their conformation upon crystallization or on dissolution of a crystal. The protein crystals examined by X-ray diffraction contain 40—50% water, thus the wet crystal is similar to a concentrated solution.

5. Conformational Transitions and Cotton Effects in the Near Ultraviolet and Visible Spectrum

Cotton effects have been observed not only in the far ultraviolet but also in the near ultraviolet (250 to 400 nm) and visible spectra of proteins. Many colorless proteins exhibit Cotton effects in the near ultraviolet, usually at 260 to 290 nm. BLOUT (1964) has suggested classifying all Cotton effects as *intrinsic* or *extrinsic* depending on whether they are related to the polypeptide backbone conformation or to prosthetic groups. Heme is such a prosthetic group in the red colored myoglobins, hemoglobins, and ferricytochromes, and they all have several absorption bands in the visible and near ultraviolet zones. Most important of these bands is that observed near 410 nm, the so called Soret band.

The rotatory dispersion of myoglobin at the Soret band is illustrated in Fig. 38 (HARRISON and BLOUT, 1965). The Cotton effect is positive and very *weak*, if com-

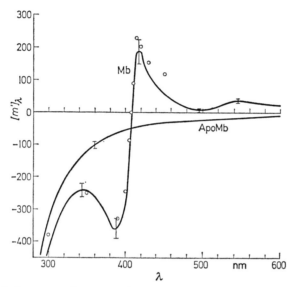

Fig. 38. The optical rotatory dispersion of myoglobin (Mb) and apomyoglobin (ApoMb) in the visible and near ultraviolet spectrum. (From HARRISON and BLOUT, 1965)

pared to the intrinsic Cotton effects in the far ultraviolet. (The magnitude of the positive extremum is only about 200°.) If heme is removed from myoglobin, this Cotton effect disappears (Fig. 38). The effect of various ligands on the rotatory dispersion of myoglobins and hemoglobins has been studied by many, *e.g.* by BEYCHOK (1964). It was found that the weak extrinsic Cotton effects were affected by the various ligands, *e.g.* the positive extremum of the Soret band Cotton effect was strongly enhanced by converting hemoglobin to CO-hemoglobin (carbonylhemoglobin).

The intrinsic and extrinsic Cotton effects of the *ferricytochromes c* have been studied in detail, *e.g.* by MIRSKY and GEORGE (1966, 1967). Fig. 39 shows the Cotton effects of the ferricytochromes c from various species in the wide spectral zone between 190 and 660 nm (from MYER and HARBURY, 1965). In the left part of the illustration the intrinsic Cotton effects of the far-ultraviolet zone are shown, while the right part shows the Cotton effects in the near-ultraviolet and visible spectrum on a strongly expanded ordinate. The differences between the cytochromes from various species are small. The magnitude of the positive extremum at the Soret band near 410 nm is approximately $+350°$ in all cases, but more significant differences are observed in the weak Cotton effects in the near ultraviolet spectrum. The far ultraviolet Cotton effect curves are typical for proteins which have a moderate helix content. ULMER (1966) has found that an 11-amino-acid heme peptide obtained in peptic digestion of the cytochrome displayed rotatory dispersion curves similar to those of the complete macromolecule; however, the curve of the peptide in the near ultraviolet was simpler than that of the intact protein. This supports the contention that the weak Cotton effects observed near the absorption maxima of he aromatic amino acids are caused

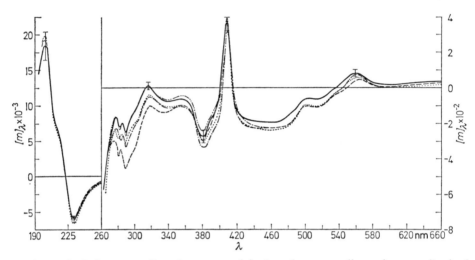

Fig. 39. Optical rotatory dispersion curves of ferricytochromes c; all samples were dissolved in 0.1 M phosphate of pH 6.8 and measured at 25°. The continuous curve shows the rotatory dispersion of horse heart ferricytochrome c, while the dashed curves represent cytochromes c from other species. (From MYER and HARBURY, 1965)

by the R-groups of these amino acids, because the 11-amino-acid heme peptide did not contain any aromatic acid residues. A theoretical analysis of the electronic transitions involved in generating the Soret band Cotton effect has been contributed by URRY (1965).

There are many blue and green proteins which have absorption bands near 600 to 650 nm and in other parts of the spectrum as well. For example, multiple Cotton effects are displayed by the *phycocyanins* isolated from blue-green algae (BOUCHER, CRESPI, and KATZ, 1966). The blue copper-containing proteins, *e.g.* the blue protein from *Pseudomonas* shows interesting Cotton effects in the visible and near-ultraviolet parts of the spectrum (TANG and COLEMAN, 1967). Removal of the copper ion abolished the Cotton effects in the visible spectrum, whereas the changes in the other parts of the spectrum were minor.

Many of the colorless proteins display weak Cotton effects in the near-ultraviolet, especially at 270 to 300 nm. A weak positive Cotton effect of lysozyme at 280 nm has been described by GLAZER and SIMMONS (1965). They found that this effect disappeared on exposure of the enzyme to sodium dodecyl sulfate, the treatment resulting in suppression of the enzyme activity. Since the magnitude of the trough at 233 nm was not affected by the treatment with detergent, it was concluded that the enzyme activity depends chiefly on the spatial arrangement of the tryptophan (and possibly tyrosine) *R*-groups which cause the 280 nm Cotton effect.

A very complex system of weak Cotton effects has been observed in *carbonic anhydrases* by MYERS and EDSALL (1965). At least six reproducible small peaks were observed in the rotatory dispersion curves between 260 and 300 nm. These weak Cotton effects disappeared when the enzyme was denatured with acid or concentrated guanidine hydrochloride solutions. The Cotton effects in this spectral zone are believed to be due to the aromatic residues of the amino acids.

Important information has been obtained from chiroptical studies on *coenzyme-apoenzyme interactions* in the presence or absence of substrates. The pyridoxal enzymes have been studied by TORCHINSKII and KORENEVA (1964), FASELLA and HAMMES (1965), and WILSON and MEISTER (1966). The last authors investigated the Cotton effects of *aspartate β-decarboxylase*. In the far ultraviolet, the curves of the apoenzyme and the holoenzyme (*i.e.* apoenzyme + pyridoxal 5'-phosphate) were very similar: both had a deep trough at 233 nm (specific rotation approximately −7000°). However, in the near ultraviolet the holoenzyme exhibited a weak Cotton effect with an inflection point at 358 nm, whereas the curve of the apoenzyme was smooth. The weak Cotton effect of the holoenzyme is a typical extrinsic Cotton effect. The work has shown that optical rotatory dispersion can serve in quantitive studies of the binding of pyridoxal 5'-phosphate to the apoenzyme. The important studies on the extrinsic Cotton effects of many heme proteins, other metalloproteins, protein–dye complexes, cofactor–apoenzyme interactions, and interactions between enzymes and substrates and inhibitors have been reviewed by ULMER and VALLEE (1965). This also includes the original work of these authors on liver alcohol dehydrogenase. It was found that upon addition of the coenzyme, reduced diphosphopyridine nucleotide (DPNH), an *extrinsic* negative Cotton effect was formed at 327 nm. This resulted in a change of the rotatory dispersion curve in the visible spectrum including an increase of levorotation at the sodium *D* line from −8° of the apoenzyme to −15° for the holoenzyme. However, the trough of the *intrinsic* Cotton effect (inflection point 225 nm, minimum at 233 to 235 nm) was not affected by the DPNH. From these observations ULMER and VALLEE (1965) concluded that addition of DPNH and formation of the extrinsic Cotton effect were not associated with alteration of the secondary or tertiary structure of the whole macromolecule, but that only a few amino acid residues on the surface of the macromolecule were involved in these asymmetry changes. It was shown that maximal trough and peak values of the extrinsic Cotton effect were observed when two molecules of DPNH combined with one molecule of the apoenzyme. Thus a new method, called *optical rotatory dispersion titration,* was introduced. In addition, these findings clearly showed that the helix content cannot be reliably estimated from optical activity measurement at sodium *D* line or from rotatory dispersion measurement in the visible spectrum.

Many more recent contributions should also be mentioned. The optical rotatory dispersion in relation to conformation of glycogen phosphorylase was investigated by JOHNSON and GRAVES (1966). WELLNER (1966) reported results on conformational changes of L-amino acid oxidase, and AKI et al. (1966) published a paper on optical rotatory dispersion and circular dichroism of D-amino acid oxidase (from pig kidney). The Cotton effects of ferredoxin and xanthine oxidase in the visible and near-ultraviolet zones have been described by GARBETT et al. (1967). Several papers on conformational changes of glutamic dehydrogenase have recently appeared; *e.g.* BAYLEY and RADDA (1966) used rotatory dispersion in a study of allosteric effects. LISTOWSKY et al. (1965) and SHIBATA and KRONMAN (1967) have studied the optical rotatory dispersion of *glyceraldehyde-3-phosphate dehydrogenase*. LISTOWSKY et al. (1965) observed a small Cotton effect in the 280 nm region in the rotatory dispersion curve of the native enzyme and this effect was diminished in the apoenzyme. Several enzyme-coenzyme systems have been studied by BOLOTINA et al. (1966, 1967). These optical rotatory dispersion studies included the effects of coenzyme binding as well as effects of substrates and inhibitors. From their measurements in the near-ultraviolet and visible spectrum, the Russian biochemists concluded that the enzyme activity of glyceraldehyde-3-phosphate dehydrogenase depended on the conformation of the apoenzyme and apoenzyme-coenzyme complex. The Moffitt constant (b_0) of native glyceraldehyde-3-phosphate dehydrogenase from hog muscle was -197, whereas the enzyme-substrate complex had a less negative b_0 of -136, and the apoenzyme had a b_0 of -131. Thus a partial disorganization seemed to occur on substrate binding, as well as on removal of the coenzyme. Unfortunately, the far ultraviolet Cotton effects were not investigated. Lactic acid dehydrogenase was studied by BOLOTINA et al. (1967) at various pH values, and it was shown that the highest enzyme activity corresponded to the highest degree of secondary structural order. Between pH 3.6 and 10.7 the changes in enzyme activity and dispersion constants were reversible.

More recently, STRICKLAND (1968) measured circular dichroism of horseradish *peroxidase* and its enzyme-substrate compounds. He found that peroxidase had positive circular dichroism bands at 190 and 406 nm, and negative bands at 207, 222, 282, and 370 nm. The far-ultraviolet bands at 190, 207, and 222 nm indicated the presence of the α-helix, and the amplitudes of the Cotton effects were compatible with a helix content of about 40%. It was found that these intrinsic Cotton effects did not change upon binding of the substrate (H_2O_2). However, significant shifts in the Cotton effects in the Soret band region and near 280 nm were observed upon formation of the peroxidase–H_2O_2 compounds. It was concluded that the helix content was not affected by the substrate binding and that the Cotton effect changes in the near ultraviolet and Soret region were caused either by some change in the orientation of the aromatic amino acid R-groups or by heme.

6. The Fine Structure of the Near Ultraviolet CD Bands and Conformational Features of the Aromatic Chromophores

As indicated before (Chapter II, Sect. 6, and Chapter IV, Sect. 5), the CD spectra in the near ultraviolet zone reveal more details on the side chain chromophores than does ORD. The important recent studies on the fine structure of absorption and CD spectra at low temperatures (Chapter IV, Sect. 5) have been extended to several proteins and in some cases it was possible to correlate the peaks with the respective chromophores.

The aromatic chromophores in *carboxypeptidase A* have been investigated with the low-temperature technique by FRETTO and STRICKLAND (1971). A series of peaks appeared in the absorption spectrum when the enzyme was cooled to − 196°. Trypto-phan maxima could be identified at 302, 295, 290.5, 283.5, 280, and 275 nm; the tyrosine had peaks at 277 nm which overlapped with tryptophan at 283.5 nm, while the phenylalanine band maxima were resolved at 268, 264, and 258 nm. The circular dichroism spectra at 24° and − 196° are shown in Fig. 40. Positive CD bands are ob-served at the shorter wavelength end of the near ultraviolet zone and negative bands were found at about 280—300 nm. Cooling brought out a strong negative CD band at 279 nm and a weak shoulder at 299 nm, while the 294 and 287 nm bands were

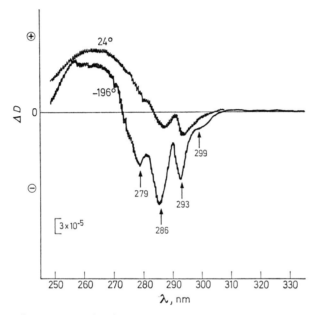

Fig. 40. Tracing of CD spectra of carboxypeptidase A dissolved in water-glycerol with 25 mM Tris-HCl + 0.4 M NaCl (pH 7.5) at 24° and −196° (after FRETTO and STRICKLAND, 1971). The negative bands centered at 286, 293, and 299 nm are assigned to tryptophan residues in asymmetric environment

intensified, sharpened and shifted slightly toward shorter wavelengths. According to FRETTO and STRICKLAND (1971), the 299, 294, and 287 nm bands arise from trypto-phan chromophores in asymmetric microenvironment. Intensification and sharper resolution on cooling is explained as due to lowering the freedom of motion of the residues, which are not quite buried in the interior of the macromolecule. Cooling shifts some of these groups into conformational states of lowest energy. In another paper, the same authors studied the effects of substrates and inhibitors on the CD spec-tra of this enzyme (FRETTO and STRICKLAND, 1971 a).

The fine structure of the CD spectrum of *insulin* was investigated by MENENDEZ and HERSKOVITS (1970). This paper is of interest for two reasons: first, as the rotatory dispersion in the near-ultraviolet range was also investigated, it showed the superior resolving power of CD; second, as insulin does not contain any tryptophan, it was easier to identify the tyrosine chromophore effects. Only negative CD bands were observed in the 230—330 nm spectral zone, and the identification of the peaks was facilitated by comparative measurements on simple model substances, *i.e.* tyrosine and phenylalanine derivatives. Zn–insulin exhibited the strongest bands in the vicinity of 260—280 nm and removal of Zn resulted in a decrease of the band strength and change in the fine structure. The effects in the 272—285 nm region were identified as being due to tyrosine *R*-groups and those at 238—267 nm as being caused by phenyl-alanine residues. As expected, no peaks were observed at 290—310 nm. Also it was shown that insulin fragments, *e.g.* the B-chain heptapeptide, yielded much weaker CD effects than the complete insulin. The insulin case is simpler because of the absence of tryptophan, but the interpretation is more difficult because of the relatively large proportion of the optically active disulfide bonds. The CD effects of the disulfide chromophores are displayed at 250—260 nm and the effects depend on the dihedral angles of the asymmetric structures.

The near-ultraviolet CD spectra of *lysozyme* have been studied in several labora-tories, also in the presence of substrate, but the interpretation of the fine structure is far from complete. The spectrum has three positive bands in the 280—300 nm zone and several poorly resolved negative bands at 250—275 nm. The latter probably arise from the disulfide, phenylalanine, and tyrosine chromophores. The positive peak centered at 283 nm may be due to either tyrosine or tryptophan *R*-groups, and the positive peaks at 288—290 nm and 294—295 nm are related to the tryptophan chromophores (TEICHBERG, KAY, and SHARON, 1970; IKEDA and HAMAGUCHI, 1972). Moreover, IKEDA and HAMAGUCHI discovered a tryptophan band at 305 nm. Con-sidering the amino acid sequences, it appears that Trp-62, Trp-63, and Trp-108 are at the active site in the macromolecule among other residues. Changes in the tryptophan bands have been observed when a substrate or inhibitor interacted with the enzyme, and the dislocation of the Trp-108 has been established as the major cause of these spectral changes. According to IKEDA and HAMAGUCHI (1972), the weak negative 305 nm tryptophan band is very sensitive to various additives, such as the substrates di- and tri-N-acetylglucosamine, which greatly enhance the 305 nm band.

Denaturation of enzymes and other highly ordered globular proteins leads to diminution of the CD bands in the near-ultraviolet zone. This is understandable, because in the unfolded state the *R*-groups, which were strongly restricted in the native form, have more freedom to attain various positions. However, in some cases denaturation brings out fine structure which was not observed in the spectrum of the native protein. STRICKLAND, KAY, and SHANNON (1970) observed this on one of the horseradish peroxidase isoenzymes. 8 *M* guanidine hydrochloride diminished the CD bands assigned to tyrosyl and tryptophanyl side chains but fine structure appeared at 268 and 261 nm, *i.e.* in the phenylalanine band zone. The intensities of these bands in the denatured peroxidase were comparable to those of an equivalent concentration of unoriented phenylalanine model compounds. According to STRICKLAND, KAY, and SHANNON (1970), the absence of the fine structure in the native enzyme is due to extensive *cancellation* of the positive and negative CD contributions of the many phenylalanine residues in different asymmetric environments. In the denatured per-oxidase, even though the unoriented residues have weak CD, all have the same CD, thereby precluding cancellation. In partial disorganization, unequal cancellation of positive and negative bands may even lead to enhancement of some near-ultraviolet peaks.

Chapter VIII

Cotton Effects and Conformation
of Nonhelical Proteins

1. Globular Proteins of Known and Unknown Conformation

A few remarks must be made regarding the terminology. First, the term "non-helical" is used for convenience only; in several cases, a very low helix content has been found by X-ray diffraction beside the β structures and nonperiodic arrangements. Second, the use of the term "globular" does not mean that the macromolecules are ideal spheres. In reality, their shapes differ from spherical, e.g. the lysozyme molecules are kidney-shaped. Highly asymmetric shapes can be detected by various methods, such as viscosity, birefringence of flow, light scattering etc. [TANFORD, 1961; JIRGENSONS 1962 (3)].

The far-ultraviolet ORD Cotton effects of some nonhelical proteins are presented in Figs. 41 and 42. Curve 1 of Fig. 41 shows the far-ultraviolet rotatory dispersion of ribonuclease, and curve 2 represents pepsin [JIRGENSONS, 1967 (1)]. In Fig. 42, curve 1 shows the rotatory dispersion of native β casein, and curves 2, 3, and 4 show the rotatory dispersion of same protein after its conformation was changed by changing the solvent (curves 2 and 4) or by treating the casein with the detergent sodium tetradecyl sulfate (curve 3) [JIRGENSONS, 1967 (2)]. Inspection of Figs. 41 and 42 shows that none of the three proteins, when in the native state, can have a significant content of α-helix. Neither ribonuclease nor pepsin (Fig. 41) displayed curves with a trough at 232 to 233 nm. Instead, a trough was observed at 225 to 230 nm. The positive peak is not at 198 to 199 but at 198 to 206 nm. The complete structure of ribonuclease is now known from X-ray diffraction studies; it will be discussed in some detail in Section 3. The conformation of pepsin is unknown, but the Cotton effects of Fig. 41 indicate that the conformations of pepsin and ribonuclease must be similar, i.e. that both probably have a significant amount of the β form. Native β casein, according to curve 1 of Fig. 42, is a random coil (disordered) protein, although other interpretations have been invoked (GARNIER, 1966). In treatments of the casein with detergent or propanol, the formation of helical and β structures is likely, as indicated by curves 2 to 4 of Fig. 42. The maxima at 198 to 200 nm are characteristic for the helix, and the maxima at 202 to 208 nm are indicative of the β-form.

In Table 14 are compiled the ORD Cotton effect data of the nonhelical proteins. The Moffitt constants, b_0, are included for comparison.

The proteins listed in Table 14 have several common characteristics: (1) their Cotton effects in the far-ultraviolet are weak; (2) their Moffitt constants (b_0), especially if computed by using a high λ_0 of 216 to 220, are close to zero; and (3) the rotatory dispersion in some cases changes on denaturation in a way that indicates that

Table 14. Far ultraviolet Cotton (ORD) effects and Moffitt constants (b_0) of nonhelical proteins in aqueous buffer solutions or water

Protein	b_0 deg.	λ_0 nm	λ_{min} nm	$[m']_{min}$ degr.	λ_{max} nm	$[m']_{max}$ degr.	References
Alcohol dehydrogenase, yeast, pH 7.3	−124	216	228	− 3,500	203	+ 8,600	Jirgensons [1966 (2)]
Do. +0.004 M Na-dodecyl sulfate	−186	216	230	− 4,900	199	+17,800	Jirgensons (1966)
Aspergillopeptidase, pH 5.5	0	212	220	− 2,400	203	+ 900	Ichishima, Yoshida (1967)
Avidin	+146	212	220	− 4,000	233	+ 1,300	Green, Melamed (1966)
Bence Jones protein, type K	+ 14	216	224	− 1,400	202—210	+ 4,700	Jirgensons, Saine, Ross (1966)
Bence Jones protein, type L	− 25	216	230	− 1,700	202—210	+ 3,800	Jirgensons, Saine, Ross (1966)
Carbonic anhydrase B, pH 7.1	+ 26	212	223	− 2,320	207	− 300	Edsall et al. (1966); Beychok et al. (1966)
Do., acid denatured, pH 1.8	− 56	212	229	− 3,570	198	+15,000	Edsall et al. (1966); Beychok et al. (1966)
Carbonic anhydrase C, pH 7.1	+ 3	212	225	− 2,900	204	+ 2,750	Edsall et al. (1966); Beychok et al. (1966)
β-Casein, pH 7.6	0	216	208	− 9,000	192	+13,000	Jirgensons [1967 (2)]
α-Chymotrypsin, pH 3.2—4.5	− 23	220	233	− 3,500	200—205	+12,400	Jirgensons (1965) and unpublished data
			215—220	− 2,700			
Chymotrypsinogen, pH 3.0—5.0	+ 6	216	233	− 3,300	229	− 3,100	Jirgensons [1967 (2)] and unpubl.
			221	− 5,000	191	+14,000	
Do., pH 3.0	− 86	212	231.5	− 3,300	228	+14,000	Smillie, Enenkel, Kay (1966)
			222	− 3,700			
Cobrotoxin, pH 5.9					233	+ 1,250	Yang et al. (1967)
Concanavalin A	− 72	212	229	− 1,200			Zand, Agrawal, Goldstein (1971)
Deoxyribonuclease, pH 4.0—5.5	− 77	216	229	− 3,800	202	+12,500	Jirgensons [1966 (2)]

Table 14 (cont.)

Protein	b_0 deg.	λ_0 nm	λ_{min} nm	$[m']_{min}$ degr.	λ_{max} nm	$[m']_{max}$ degr.	References
Do. +0.01 M Na,dodecyl sulfate	−156	216	231	− 4,300	198	+19,700	Jirgensons [1966 (2)]
Elastase, pH 7.9	+ 26	220	210—214	− 3,000			Jirgensons (1965)
Fetuin, pH 7.8	−125	205	229	− 3,100			Verpoorte and Kay (1966)
Globulin, IgG, human, pH 6—8	+ 36	216	221	− 2,100	210	+ 5,200	Jirgensons [1966 (1)]
Glucose oxidase, pH 7.2	− 56	220	228	− 3,300	202	+18,000	Jirgensons (1965)
Kallikrein inhibitor	−169	212	225	− 6,000	195	+10,000	Scholtan and Rosenkranz (1966)
β-Lactoglobulin	− 60	212	237 / 230	− 1,500 to − 2,300	203	+ 8,500	Sarkar, Doty (1966); Timasheff, Townend, Mescanti (1966)
Luteinizing hormone, pH 4—8	+ 18	216	220—228	− 3,000			Jirgensons [1966 (3)]
Lysostaphin, pH 4.0			240	− 380	231	− 140	Trayer and Buckley (1970)
Myeloma globulin IgG, pH 7.0	+ 23	216	228 / 199	− 1,950 / − 7,400	210 / 205	+ 3,500 / + 3,700	Ross (1967); Ross and Jirgensons (1968)
Myeloma globulin IgG, pH 7.0			230 / 226	− 1,600 / − 1,500			Dorrington, Zarlengo, Tanford (1967)
Pepsin, pH 5.0—5.2	+ 8	216	225—228	− 4,800	202—206	+ 7,100	Jirgensons [1966 (3)]
Do. pH 5.3	− 14	216	226	− 4,800	191	+17,000	Jirgensons [1967 (1)]
Phosvitin, in water	+ 15	216	208	−14,000			Jirgensons [1966 (1)]
Ribonuclease, in water			228	− 4,200	198—206	+ 9,200	Jirgensons [1967 (1)]
Trypsin, pH 2.7—3.5	− 19	216	224—228	− 2,500 to − 3,100	203	+ 3,500	Jirgensons [1966 (1)]
Trypsin inhibitor, soybean, pH 3.1	+ 35	216	209	− 7,700	193	+11,200	Jirgensons [1966 (2)]

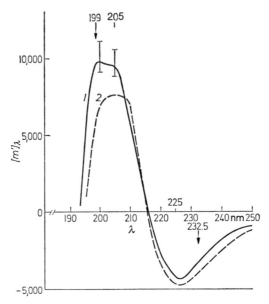

Fig. 41. The far ultraviolet rotatory dispersion of aqueous solutions of ribonuclease (curve *1*) and pepsin (curve *2*). [JIRGENSONS, 1967 (1)]

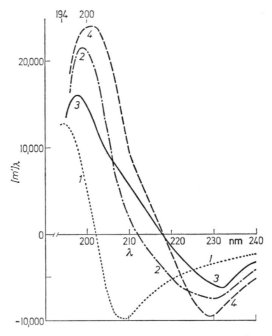

Fig. 42. The Cotton effects of β-casein. Curve *1*, casein in 0.03 M sodium phosphate, pH 7.6. Curve *2*, casein in 50% n-propyl alcohol (v/v), pH 7.2. Curve *3*, casein with 0.0012 M aqueous sodium tetradecyl sulfate, pH 2.2; the solution was kept at 75° for 1 h and cooled. Curve *4*, casein in 90% propanol, pH 2.8. The concentration of the β-casein was 0.02%; the optical path was 0.050 to 0.50 cm. [JIRGENSONS, 1967 (2)]

the helix content is increasing instead of decreasing. *Avidin* is one of the strangest examples in that it has a positive b_0, and the Cotton effect curves have a trough at 220 nm and a peak at 233 nm in the positive part of the chart (GREEN and MELAMED, 1966). YANG et al. (1967) have recently reported a similar case, *i.e. cobrotoxin*, which has a positive peak at 233 nm of the same magnitude ($+1250°$) as avidin. YANG et al. concluded that the positive peak at 233 nm might be considered an indication of the presence of the *left-handed helix* a proposition that should be considered with caution. In all the other cases where a trough is observed at 220 to 230 nm and a positive peak at 200 to 210 nm the presence of the β structure can be suspected.

From 1968 on, more attention was devoted to circular dichroism, and many proteins were studied by this technique. Table 15 shows the far-ultraviolet circular dichroism band maxima and mean residual molar ellipticities of a group of nonhelical proteins. Some of the cases will be discussed in the next sections, especially in relation to the results obtained in X-ray structural analysis.

2. Globular Proteins having a High Content of β Conformation

X-ray diffraction of monocrystals and infrared spectroscopy are the two independent methods which have been used to detect the structural orders in proteins and to ascertain the presence of the β conformation. X-ray diffraction has disclosed a high amount of the extended β structure in α-chymotrypsin (MATTHEWS et al., 1967), chymotrypsinogen, carbonic anhydrase, elastase, and trypsin. Infrared spectroscopy has indicated considerable amounts of the β structure in β-lactoglobulin (TIMASHEFF and SUSI, 1966) and in deoxyribonuclease (CHENG, 1966). The latter author concluded that about 50% of the chains of deoxyribonuclease have the β structure and that the rest are unordered. In fair agreement with our findings, he found in the ORD curves of this protein a trough at 228 nm and a positive peak at 205 nm.

The conformation of polypeptide chains in α-*chymotrypsin*, according to X-ray diffraction data, is schematically shown in Fig. 43 (MATTHEWS et al., 1967). "With the exception of eight residues at the C-terminus which form a short section of α-helix, the chains tend to be fully extended. As these chains trace through the molecule they often run parallel to one another at a distance of 5.5 to 5 Å for stretches of several residues. In a fully extended polypeptide chain the peptide groups are available for hydrogen bonding, and therefore the tertiary structure is presumably stabilized to a large extent by hydrogen bonds between the chains. There is a very noticeable tendency for a chain to interact with a neighbouring region in the sequence. This occurs frequently when a chain turns back and runs parallel to its previous course. The overall appearance is of a chain 'piled on itself' rather than rolled, coiled, or knotted. This suggests that the molecule might be folded around nuclei of highly stabilized local conformation" (MATTHEWS et al., 1967). X-ray diffraction has thus revealed that in chymotrypsin the β structures predominate and that the helix content is negligible. The X-ray diffraction of the crystals of the zymogen — α-*chymotrypsinogen* — has also been studied (KRAUT, 1971).

The ORD of α-chymotrypsin and α-chymotrypsinogen have been studied in several laboratories (RAVAL and SCHELLMAN, 1965; BILTONEN et al., 1965; FASMAN, FOSTER, and BEYCHOK, 1966; SMILLIE, ENENKEL, and KAY, 1966; and the work of the writer).

The far-ultraviolet ellipticity curves of chymotrypsin and chymotrypsinogen are shown in Fig. 44. The curves do not have the negative trough at 222 nm which is typical for the α-helix. This is in agreement with the X-ray structural analysis which revealed only a very short helix at the C-terminal end of the backbone in both proteins. In Fig. 44, the chymotrypsinogen curve is resolved into Gaussian components yielding a large negative band centered at 213—214 nm. This is also in agreement

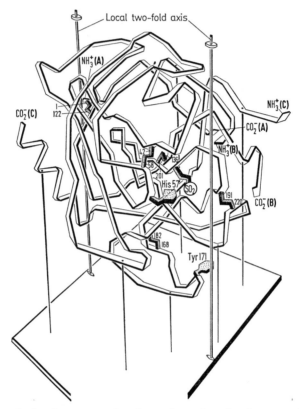

Fig. 43. A schematic drawing representing the conformation of α-chymotrypsin, according to X-ray diffraction studies of MATTHEWS et al. (1967). The disulfide bridges, the side groups of histidine-57, tyrosine-171, and the sulfonylated serine-195 are shown. Also, the N and C terminal groups of the A, B, and C chains are indicated. Many chain segments run parallel to each other, and helical conformation is present only at the C terminal of the C chain

with the X-ray diffraction data (KRAUT, 1971) that indicated a high content of the extended chains in the pleated-sheet conformation. The other bands, however, are not so easy to identify. The negative bands at 228—235 nm probably should be related to the aromatic R-groups. STRICKLAND, HORWITZ, and BILLUPS (1969) have investigated in detail the tryptophan bands of chymotrypsinogen which appear in the 290—305 nm zone by comparing the circular dichroism of this protein and some simple tryptophan derivatives at various temperatures. They found that at 77° K the rotatory strength of N-acetyl-L-tryptophanamide is 18 times greater than at 298° K, which indicates a high degree of conformational mobility at room temperature. In contrast, chymotrypsinogen showed only 25⁰/₀ intensification of the CD bands upon cooling to 77° K. From this it was concluded that the tryptophan residues within the macromolecule of chymotrypsinogen have relatively rigid positions at room temperature (Chapter IV, Sect. 5).

The ORD effects of *trypsin* have been studied also by D'ALBIS (1966). In accord with our work [JIRGENSONS, 1966 (1)], a negative extremum was found at 228 nm.

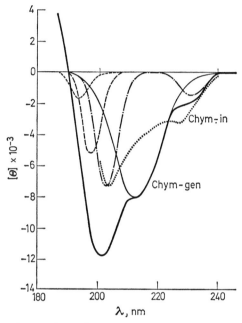

Fig. 44. The far-ultraviolet circular dichroism spectra of chymotrypsin and chymotrypsinogen. The ellipticity curve of chymotrypsinogen is resolved into Gaussian components; the major component is centered at 214 nm

and the magnitude of the extremum, depending on the acidity, was found to be between −2500 and −3600° (uncorrected). The major conformational order in trypsin, according to D'ALBIS, is the β structure. The measurements, however, were not extended to the very short wavelength zone.

The far-ultraviolet CD spectra of trypsin (PANTALONI, D'ALBIS, and DESSEN, 1968; TAMBURRO, SCATTURIN, and ROCCHI, 1968; JIRGENSONS, 1970) and elastase (JIRGENSONS, 1970; VISSER and BLOUT, 1971) are of the same type as those of chymotrypsin and chymotrypsinogen, thus indicative of the presence of pleated-sheet conformation. A very low helix content and a significant amount of pleated-sheet conformation indeed has now been confirmed in trypsin (STROUD, KAY, and DICKERSON, 1972) and in elastase (HARTLEY and SHOTTON, 1971) by high-resolution X-ray structural analysis.

Table 15. Far-ultraviolet circular dichroism spectra of nonhelical proteins

Protein	pH	Band maxima at nm	$[\Theta]$ deg. cm²/ decimole	References
Acetoacetate decarboxylase	5.9	219	− 8,400	Lederer (1968)
		198	+ 7,100	
L-Asparaginase, L5, *E. coli*	7.4	215	− 7,000	Rosenkranz and
		195	+10,000	Scholtan (1971)
Bence Jones protein ϰ-type	7.0	217	− 3,000	Kincaid and
		202	+ 3,900	Jirgensons (1972)
do., λ-type	7.0	217	− 4,500	Kincaid and
		201	+ 5,800	Jirgensons (1972)
		193	+ 2,700	
α-Casein B	7.5	213—217	− 4,700	Timasheff et al. (1967)
		200	− 8,300	
Ceruloplasmin	6.2	219	− 3,900	Hibino et al. (1969)
		199	+ 4,300	
α-Chymotrypsin	5.7	229	− 3,600	Hunt and Jirgensons
		202	− 7,400	(unpubl.)
α-Chymotrypsinogen	3.0	230	− 2,400	Gorbunoff (1969)
		205	−12,000	
		200	−13,000	
Concanavalin A	5.2	218—224	− 7,800	Kay (1970)
		197	+13,200	
α-Crystallin	8	216	− 3,000	Li and Spector (1967)
γ-Crystallin, dogfish	2—10	233	+ 400	Jones and Lerman
		217	− 4,000	(1971)
		208	− 5,000	
		193	+12,000	
Deoxyribonuclease, pancreatic	3.4	214—218	− 7,000	Jirgensons (1970)
		210	− 6,000	
		194	+11,600	
Elastase	4—6	234	− 800	Visser and Blout
		219	− 2,700	(1971); Jirgensons
		205	− 7,600	(unpubl.)
		198	− 9,900	
Immunoglobulin, IgG, human	7.0	217	− 2,500 [a]	Doi and Jirgensons
		202	+ 2,600 [a]	(1970)
β-Lactoglobulin	5	215	− 5,500	Townend,
		196	+ 8,400	Kumosinski, and Timasheff (1967)
Luteinizing hormone, LH	6—7.7	235	− 320	Jirgensons and Ward
		205—215	− 8,000	(1970)
Pepsin	4.3	212—218	− 8,000	Funatsu et al. (1971)
		194	+ 6,000	
Pepsin	5.7	212	− 6,000 [a]	Nakagawa and
		195	+ 5,000 [a]	Perlmann (1970)
Pepsinogen	7.7	212	− 6,000 [a]	Nakagawa and
		195	+ 5,400 [a]	Perlmann (1970)
Phosvitin	3.4	197	−18,000	Taborsky (1968)
	1.8	215	−11,800	
		194	+14,600	

Table 15 (cont.)

Protein	pH	Band maxima at nm	$[\Theta]$ deg. cm²/ decimole	References
Ribonuclease	6.8	215—220 (shoulder)		TAMBURRO, SCATTURIN,
		208	−11,000	and MORODER
		196	+ 6,000	(1968)
Trypsin	3.0—6.5	218—222 (shoulder)		JIRGENSONS (1970)
		207	− 4,700	
		200	− 3,800	
Trypsin inhibitor, from lima beans	7.5	202	−11,000	GORBUNOFF (1970)
Trypsin inhibitor, from soya beans	7.3	226	+ 1,570	JIRGENSONS,
		200	−17,500	KAWABATA, and
				CAPETILLO (1969)

[a] These data were corrected for the refractive index of the solvent.

Inspection of Table 15 suggests the presence of pleated-sheet structures in many other proteins, particularly in those which have negative CD bands at 210—220 nm and positive peaks at 193—205 nm. Deoxyribonuclease is one such candidate, as mentioned before. According to TIMASHEFF and BERNARDI (1970), the pancreatic deoxyribonuclease and acid deoxyribonuclease from spleen are rich in pleated-sheet conformation, whereas deoxyribonuclease from *Staphylococcus aureus* has about 20—25% helix, as indicated also by TANIUCHI and ANFINSEN (1968) who studied this enzyme in great detail.

3. Cotton Effects and Conformation of Bovine Pancreatic Ribonuclease

The conformation of *ribonuclease* has been studied by X-ray diffraction by several groups of investigators, notably KARTHA, BELLO, and HARKER (1967) and WYCKOFF et al. (1967). A schematic diagram of the main chain folding is shown in Fig. 45 (KARTHA, BELLO, and HARKER, 1967). The model was derived from 2 Å electron density maps involving 7,294 reflections and data from seven crystalline ribonuclease derivatives containing heavy atoms. The macromolecule appears to be roughly kidney shaped with dimensions of $38 \times 28 \times 22$ Å, and has a deep depression in the middle of one side. Since the complete amino acid sequence (primary structure) was known from chemical structural studies, it was possible to identify the side chains at the various points of the main chain. According to the model, the secondary and tertiary structures are more irregular than in lysozyme and myoglobin. There are only a few helical twists and no perfect α-helical strands. Most of the chain is strongly extended, as in chymotrypsin, but there are fewer parallel segments than in chymotrypsin. The chain sections between amino acid residues 40 to 58 and 98 to 110 run in one direction and those of region 75 to 90 in the opposite direction, but a perfect β structure is not discernible.

WYCKOFF et al. (1967) derived a similar model from electron density maps of a lower (3.5 Å) resolution. They used ribonuclease-S, which differs from native ribonuclease in that the peptide bond between the amino acid residues 20 and 21 was cleaved with subtilisin (RICHARDS and VITHAYATHIL, 1959). The N-terminal peptide remains linked to the rest of the macromolecule by secondary (non-covalent) bonds, the enzyme remains active, and the conformation in the rest of the macromolecule is believed to be the same as in the complete native enzyme. In the model of WYCKOFF et al. (1967) the helical strands are more clearly shaped than in the model of KARTHA, BELLO, and HARKER; and the same can be said about the β structures. According to

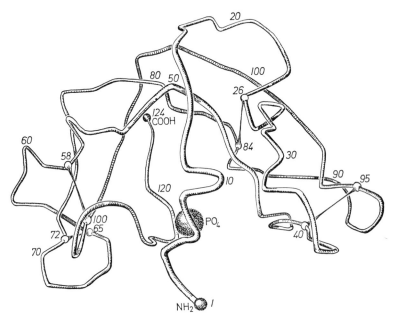

Fig. 45. Schematic drawing of the polypeptide chain in the macromolecule of ribonuclease, according to X-ray diffraction studies of KARTHA, BELLO, and HARKER (1967)

WYCKOFF et al., about 15% of the polypeptide chain of ribonuclease-S is in the helical form, and there is an appreciable antiparallel β chain pairing. According to DOSCHER and RICHARDS (1963), the enzyme activity of ribonuclease-S crystals is the same as in a solution. This can be regarded as evidence that the conformation in solution is not significantly different from the conformation in the protein crystals.

The far-ultraviolet rotatory dispersion of chromatographically pure ribonuclease-A is shown in Fig. 41, and the curve clearly indicates a low helix content and the possible presence of the β conformation. This is in agreement with the X-ray diffraction data, indicating that the macromolecules of the enzyme are not disorganized upon dissolution of the crystals. However, a quantitative estimate of the content of helix and β structure from rotatory dispersion data is not warranted for a variety of reasons. One is the rotatory contribution of the amino acid side groups; another reason is that it is not known how the imperfections of the helix and β conformations, as revealed in high-resolution X-ray diffraction, influence the Cotton effects. It may be speculated that the shift of the trough from 233 nm of the helix and 229 to 230 nm of the β form to 225 to 228 nm (Fig. 41) in the ribonuclease curve may be due to these somewhat distorted and short helices and β arrangements.

More recently, the earlier rotatory dispersion data on bovine pancreatic ribonuclease have been confirmed and extended by circular dichroism studies in several laboratories. Fig. 46 shows tracings of the CD recordings of native (curve 2) and fully unfolded and disorganized (curve 3) bovine ribonuclease, a tracing of lysozyme (curve 1) being given for comparison. The lower magnitudes of the extrema of ribo-

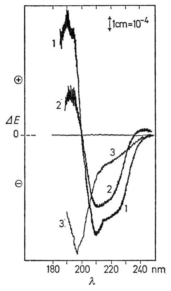

Fig. 46. Tracings of circular dichroism recordings of lysozyme (curve 1), bovine pancreatic ribonuclease in native state (curve 2) and oxidized and completely unfolded ribonuclease (curve 3) in the far-ultraviolet spectral zone. The concentrations of the proteins were 0.01%, the optical path through the solution was 0.10 cm. (JIRGENSONS, 1970)

nuclease in comparison to lysozyme, at 190—195 nm and 210—220 nm respectively, are in accord with the lower helix content of ribonuclease (curve 2 versus curve 1). The disappearance of these Cotton effects and appearance of the negative band at 196—198 nm on denaturation is typical for this group of proteins (curve 3). The chiroptical properties of this enzyme have been studied also by CATHOU, HAMMES, and SCHIMMEL (1965), SIMPSON and VALLEE (1966), and PFLUMM and BEYCHOK (1969), among others. Besides the weak positive CD band near 240 nm, weak negative bands were observed in the near-ultraviolet zone at 260—285 nm, and the latter were ascribed chiefly to the tyrosine R-groups. An extensive study of these chromophores was contributed by HORWITZ, STRICKLAND, and BILLUPS (1970) who compared the enzyme CD spectra with those of simple tyrosine derivatives at the ordinary temperature of 298° K and very low temperature of 77° K. The interpretation was simplified by the fact that ribonuclease does not contain tryptophan. However, the spectroscopic analysis leads to the conclusion that in the 275—310 nm spectral region the circular dichroism is caused not only by fixed tyrosine chromophores but also by the disulfide bonds.

One of the greatest achievements in modern chemistry is the complete step-by-step synthesis of ribonuclease A by GUTTE and MERRIFIELD (1971) with the aid of the solid-phase method. Since the 124 amino acid residues were linked by chemical methods and the synthetic macromolecule exhibited enzyme activity, there is no longer any doubt that conformation is determined by the primary structure of the polypeptide chain.

4. Other Rigid Nonhelical Proteins

Ribonuclease, chymotrypsin, chymotrypsinogen, and deoxyribonuclease are globular proteins, *i.e.* the macromolecules are not highly asymmetric. The solutions of these proteins have a low viscosity which depends only slightly on ionic strength. In spite of the fact that the macromolecules of all these proteins are globular, and that the polypeptide chains are folded into compact and rigid bodies, the conformations differ considerably.

The *soybean trypsin inhibitor* is one of the most interesting rigid nonhelical proteins. Its ORD Cotton effect curves are shown in Fig. 47. Curve *1* represents the far-ultraviolet rotatory dispersion of an aqueous solution of native soybean trypsin inhibitor, while curves *2* to *5* show the Cotton effects of the inhibitor in various water–propanol mixtures (25% n-propanol by volume for curve *2*, and 75% for curve *5*). These data indicate that native soybean trypsin inhibitor is of aperiodic conformation, and that the formation of helical and β structures is promoted by propanol. Curve *1* of Fig. 47 is very similar to the curve of β casein (curve *1*, Fig. 42) and both curves are similar to those representing fully disordered chains. However, there are large differences in the *gross conformation* in these three cases: the polyglutamate in nearly neutral solutions is in the form of flexible highly asymmetric chains, the casein particles are relatively compact random coils, and the molecules of the trypsin inhibitor are *rigid*. The difference between β-casein and soybean trypsin inhibitor is well illustrated by the difference in the *viscosity* of their solutions: the intrinsic viscosity of β-casein is 0.144 dl/g, whereas the intrinsic viscosity of soybean trypsin inhibitor is 0.03 to 0.04 dl/g, depending on pH and ionic strength (JIRGENSONS, unpublished data). These findings lead to the problem of *how the flexibility of unordered chains affects rotatory dispersion*. This problem has not been studied in detail.

With respect to amino acid composition, soybean trypsin inhibitor is a perfectly normal protein (WU and SCHERAGA, 1962), whereas β-casein has a high proline content (15.1%), which is unfavorable for the formation of both the α-helical and β structures. GARNIER (1966) concluded that β-casein might have some polyproline helix in some parts of the macromolecule. However, the proline content is low in soybean trypsin inhibitor: there are only 10 proline residues in the total of 181. Thus the deep trough in the trypsin inhibitor curve at 206 to 210 nm cannot be explained by the presence of the polyproline helix. The single polypeptide chain in the trypsin inhibitor is looped by only two disulfide bridges, and the content of the aromatic amino acids in the protein is low.

Even more interesting are the circular dichroism spectra of this inhibitor (Fig. 48). In the far-ultraviolet, this protein has a weak positive peak at 226 nm and a relatively strong negative band near 200 nm, and there are a series of weak bands in the 240—320 nm zone, *i.e.* at 295, 287, 273, 261, and 246—252 nm (IKEDA et al., 1968; JIRGENSONS, KAWABATA, and CAPETILLO, 1969). The CD spectra do not change much in acid (pH 2.2), but a drastic change was observed on addition of sodium dodecyl sulfate to the acid solution of the inhibitor (JIRGENSONS, 1972). The CD bands in the

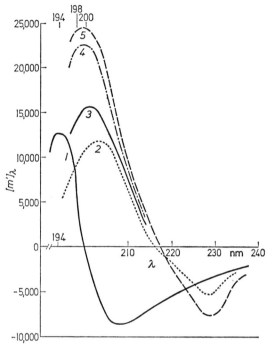

Fig. 47. The Cotton effects of aqueous solutions of soybean trypsin inhibitor, and of the protein in water-propanol mixtures. Curve *1*, native soybean trypsin inhibitor in water. Curve *2*, the same in 25% n-propanol, (v/v); curve *3*- 37.5%; curve *4*- 50%; curve *5*- 75% (v/v) n-propanol. The concentration of the protein was between 0.01 and 0.02%; the optical path was between 0.050 and 0.50 cm. [JIRGENSONS, 1967 (2)]

near-ultraviolet zone were eliminated and the spectrum in the far-ultraviolet was transformed into one characteristic for proteins of moderate α-helix content. Since soybean trypsin inhibitor is a rigid globular protein, the negative band at 200 nm apparently does not indicate the presence of random coil but rather of some constrained nonperiodic folding which is expressed by a similar CD band in the spectra of elastase, trypsin, chymotrypsin, and chymotrypsinogen (see Table 15 and Fig. 44). In the near-ultraviolet region, the negative bands at 295 and 287 nm probably arise from tryptophan chromophores and the band at 273 nm may be related to both tyrosine and tryptophan; the disulfide bonds also may be involved in the CD effects in the 250—280 nm zone. Most interesting is the observation that acid or *6 M* guanidine hydrochloride did not affect much these near-ultraviolet bands, whereas they were nearly eliminated by sodium dodecyl sulfate. Since the long hydrophobic tail of the detergent interacts with the hydrophobic regions of the macromolecule, and since it changes the conformation so effectively, the hydrophobic forces must be decisive in maintaining the globular folding of this protein.

The effects of the disulfide bridges on the CD spectra and stabilization of the native conformation also have been considered. It is known that cystine and its derivatives display a strong negative CD effect near 200 nm (COLEMAN and BLOUT, 1968). However, since only two disulfide bonds are involved among the total number

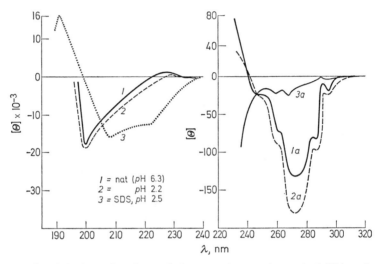

Fig. 48. Circular dichroism of native and detergent-denatured trypsin inhibitor from soya beans. Curves *1* and *1a*, native inhibitor, pH 6.3, without dodecyl sulfate. Curves *2* and *2a*, inhibitor without detergent but in acid solution, pH 2.2. Curve *3*, 0.0062% inhibitor + 0.024 *M* sodium dodecyl sulfate, pH 2.4; curve *3a*, 0.062% inhibitor + 0.24 *M* dodecyl sulfate, pH 2.4. Note the difference in the expansion scale of the ordinates in the far- and near-ultraviolet spectral zones. (JIRGENSONS, 1972, and unpubl.)

of 181 amino acid residues in the inhibitor, their effects can hardly be decisive (GORBUNOFF, 1970). The changes in the CD spectra which we observed after reduction of the disulfide bonds of the inhibitor (JIRGENSONS, KAWABATA, and CAPETILLO, 1969) may be caused chiefly by the unavoidable disorganization of the original conformation. The disulfide bond effects on the conformation and CD spectra may be more important in the trypsin inhibitor isolated from lima beans. This inhibitor contains seven disulfide bonds per total of only 80 residues, but its CD spectrum has a negative band near 200 nm as has the soybean protein (GORBUNOFF, 1970). The fact that these and several other proteins of different molecular weight, amino acid composition and different conformation act as trypsin inhibitors in a similar way indicates that only certain short amino acid sequences participate in the inhibition reaction (LASKOWSKI and SEALOCK, 1971).

Many other globular proteins belong to the same group of globular proteins, *i.e.* the nonhelical proteins the secondary structure of which is due to hydrophobic interactions and which have very little helix (usually distorted) and also little pleated sheet conformation. The luteinizing hormone, also called interstitial cell-stimulating hormone, displays CD spectra similar to those of the trypsin inhibitors described above (JIRGENSONS and WARD, 1970; PERNOLLET and GARNIER, 1971; BEWLEY, SAIRAM, and LI, 1972). This hormone is a globular protein with a high cystine content; but as it is a glycoprotein, it will be discussed in Chapter XI. The neurophysins of the posterior pituitary gland are also nonhelical, yielding CD spectra similar to the spectra of the other proteins described in this section (BRESLOW et al., 1971).

5. Cotton Effects and Conformation of Flexible Nonhelical Proteins

Phosvitin from egg yolk is an example of flexible nonhelical proteins. Moreover, like polyglutamate, it is a *linear polyelectrolyte*. Phosvitin contains 9.7% phosphorus, which is bound to the hydroxyl groups of the many serine residues in the form of monoesterified phosphate. This results in a *high negative charge density* along the

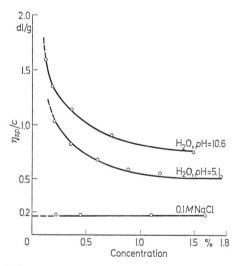

Fig. 49. Dependence of the reduced specific viscosity η_{sp}/c on concentration, c, for aqueous and alkaline solutions of phosvitin, and of phosvitin in 0.10 M NaCl. [JIRGENSONS, 1966 (1)]

polypeptide chain. Aqueous solutions of phosvitin, at low ionic strength, show the typical exponential increase of the reduced specific viscosity on dilution, as exhibited by many synthetic linear polyelectrolytes. At higher salt concentration the normal linear dependence of the reduced specific viscosity is observed, and intrinsic viscosity values of about 0.15 to 0.3 dl/g are found. This is illustrated in Fig. 49. According to these data, at low ionic strength the chain is stretched out because of electrostatic repulsion; in the presence of 0.1 M NaCl, the viscosity decreases and the chains appear as compact random coils. This viscosity behavior of phosvitin is comparable to that of disordered polyglutamate and differs profoundly from that of soybean trypsin inhibitor and other rigid proteins.

The Cotton effect curves of phosvitin are shown in Fig. 50. In the pH range of 3—10 and low ionic strength, the CD and ORD curves are in accord with the described results on viscosity. In more acidic solutions (pH 1.8), the chiroptical properties indicate a transition from disordered to pleated-sheet conformation (TABORSKY, 1968; Fig. 50). As the ionization of the phosphate groups is reduced by the hydrogen ions, the polypeptide chain is folded into a more compact unit and the formation of hydrogen bonded segments in the macromolecule is possible. A similar transition occurs on increasing the ionic strength, e.g. by adding NaCl, whereby the negative

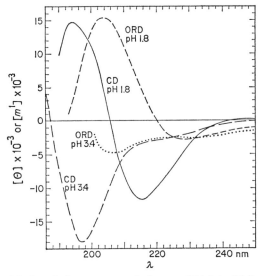

Fig. 50. CD of phosvitin in salt-free aqueous solution at pH 1.8 *(solid line)* and pH 3.4 *(long dashed line)*. For reference, the ORD of corresponding phosvitin solutions is included in the figure: pH 1.8 *(short dashed line)* and pH 3.4 *(dotted line)*. The protein concentration was 0.2 mg per ml. Each solution was kept at the indicated pH for a period of at least 1 hour before measurement. (From TABORSKY, 1968)

CD band centered near 197 nm is weakened. The ORD and CD of phosvitin were studied also by the writer and by GRIZZUTI and PERLMANN (1970, 1971).

The ordered nonhelical proteins described in previous sections, e.g. soybean trypsin inhibitor or chymotrypsin, differ in their circular dichroism behavior from phosvitin and other disordered proteins by two major features: 1) that the strong far-ultraviolet negative band of the former is observed at 200—205 nm, whereas the latter have it at 197—198 nm; and 2) that the former have a characteristic fine-structured CD spectrum in the near ultraviolet zone at 240—320 nm, whereas the latter have a very weak dichroism without any well-defined fine structure. The latter fact is explained by the absence of any rigid tertiary structure in the flexible proteins.

Several basic proteins, which in solution have the properties of positively charged flexible polyelectrolytes, have been isolated from spinal cord (CHAO and EINSTEIN, 1970) and brain (KORNGUTH and PERRIN, 1971). Their viscosity and chiroptical properties are similar to those of phosvitin. The problem of the relationship between CD and the flexibility of synthetic random-coil polyamino acids has been considered by FASMAN, HOVING, and TIMASHEFF (1970) who studied the CD of films of varying moisture content.

The *histones* represent another group of flexible random-coil proteins. In amino acid composition they differ strongly from the phosvitins in that the histones are *basic* proteins containing large amounts of lysine and arginine. The optical activity of aqueous histones is very similar to that of phosvitin, chiefly in the magnitudes and

Table 16. Rotatory dispersion and intrinsic viscosity of fully disordered proteins according to TANFORD et al. (1967) and TANFORD, KAWAHARA, and LAPANJE (1967). The proteins were dissolved in 6 M guanidine-HCl. The Moffitt constants, b_0, were determined by using a λ_0 of 212

Protein	Mol. wt.	b_0, degr. S-S intact	b_0, degr. S-S reduced	$[m']_{233}$, degr. S-S intact	Intrinsic viscosity, dl/g; S-S reduced
Ribonuclease	13,680	− 2	−5	−2350	0.166
β-Lactoglobulin	18,400	−15	0	−2600	0.228
Chymotrypsinogen	25,700	−13	−9	−2500	0.268
Pepsinogen	40,000	+ 1	+6	−2350	0.315
Serum albumin	69,000	+ 6	−7	−2450	0.522

positions of the troughs and peaks; but since there are significant differences between the various histone fractions, they will be described in more detail in Chapter X. In the presence of electrolytes, especially polyvalent anions and anionic detergents, the conformation of the histones is changed drastically.

Extensive studies on the rotatory dispersion and viscosity of many globular proteins which were *fully denatured* in 6 M guanidine hydrochloride have been reported by TANFORD, KAWAHARA and LAPANJE (1967) and by TANFORD et al. (1967). The data are compiled in Table 16. The conclusion that the proteins were fully disordered is supported by both rotatory dispersion and viscosity data: the Moffitt constants (b_0) in all cases are near zero, the $[m']_{233}$ values are between − 2350 and − 2600°, and the intrinsic viscosity is very high and increases with the molecular weight of the protein. This is a clear indication that the polypeptide chains were devoid of secondary and tertiary structure.

6. Cotton Effects and Conformation of Immunoglobulins

The nonhelical nature of serum γ-globulins was discovered in 1957/58 [JIRGENSONS, 1958 (1)]. This conclusion was based on the rotatory dispersion data obtained from measurements in the visible and near-ultraviolet zones. In 1964 spectropolarimetric techniques were considerably improved and it was possible to reach the far-ultraviolet zone of about 210 to 215 nm. According to these early findings, the human serum γ-globulins (immunoglobulins) had a trough at 228 nm, and the specific rotation at this wavelength was −2200° (JIRGENSONS, 1964). Later, with the aid of still more improved techniques, it was possible to reach the 190 to 210 nm zone, and it was found that a normal human immunoglobulin exhibited a peak at 210 nm, and that the rotatory power at this point was +3000 to about +5000° [JIRGENSONS, 1965, 1966 (1, 3)]. This finding was confirmed by SARKAR and DOTY (1966).

Since the pathological 7S human IgG immunoglobulins are known to be very homogeneous, recent Cotton effect studies have concentrated on them. In our laboratory, attention was centered chiefly on the far-ultraviolet zone of 190 to 240 nm, while others have investigated mainly the 220 to 300 nm spectral region (DORRING-

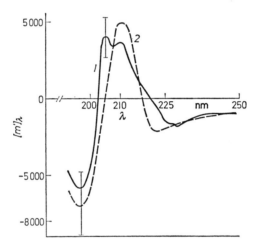

Fig. 51. Optical rotatory dispersion of immunoglobulins in the far ultraviolet. Curve *1*, a myeloma IgG 7S immunoglobulin from an individual donor. Curve *2*, pooled normal human IgG 7S immunoglobulin. The curves represent average values from many determinations using diluted solutions of the globulins in 0.01 *M* phosphate buffer of pH 7.0 to 8.2; optical path 0.050 to 0.50 cm

TON, ZARLENGO, and TANFORD, 1967). The 7S immunoglobulins are known to have a rather complex chain structure. They are composed of two heavy (H) and two light (L) chains linked by disulfide bonds and secondary linkages. The rotatory dispersion of the separated chains has also been studied recently; the same is true of some of the immunoglobulin fragments obtained in limited proteolysis. The urinary Bence Jones proteins are identical to the light chains of the serum immunoglobulins, and their Cotton effects also have been investigated.

Fig. 51 shows the far ultraviolet Cotton effect curves of a practically homogeneous 7S human myeloma IgG immunoglobulin (curve *1*) and of a normal pooled human 7S IgG immunoglobulin (curve *2*), as found in our laboratory (unpublished results). The curves have some resemblance to those of ribonuclease and other nonhelical proteins, and they differ strongly from the curves exhibited by soybean trypsin inhibitor or the flexible random-coil proteins. The immunoglobulin curves indicate that some sort of β conformation might be present, yet an independent proof of this is lacking.

The degree of precision in the case of immunoglobulins is low, as indicated by the vertical bars. For this reason, the position and magnitude of the trough at 195 to 200 nm are uncertain. There also were considerable doubts about the double peak at 203 to 212 nm, yet repeated measurements on several individual myeloma IgG specimens have confirmed the complex nature of the curve (Ross, 1967; Ross and Jirgensons, 1968; and unpublished data). Individual Bence Jones proteins also had similar double peak curves (Jirgensons, Saine, Ross, 1966).

Fig. 52 shows some of the results of Dorrington, Zarlengo, and Tanford (1967). The curves of a native myeloma IgG immunoglobulin and the globulin re-

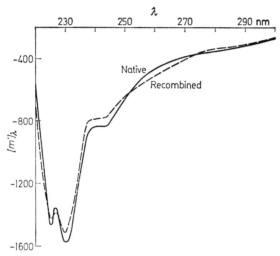

Fig. 52. Rotatory dispersion curves of native myeloma immunoglobulin and of the globulin reconstituted from separated H and L chains. (From Dorrington, Zarlengo, and Tanford, 1967)

constituted from the separated H and L chains are shown between the wavelengths of 220 and 300 nm. Both curves have two troughs at 225 to 230 nm and weak Cotton effects at the longer wavelengths. This is in agreement with observations made in our laboratory. It is remarkable that the conformation could be reformed after separation and recombination of the chains. The rotatory dispersion pattern was not much affected by rupture of interchain disulfide bonds, but it was changed when the proteins were exposed to propionic acid.

Steiner and Lowey (1966) found that the rotatory dispersion patterns of "nonspecific" purified rabbit immunoglobulin differ from the specific anti-2,4-dinitrophenyl antibody. The measurements were made in the 220 to 270 nm zone, and the curves were similar to those described previously. Moreover, Steiner and Lowey (1966) investigated the fragments obtained by digestion with papain. The immunochemically active fragment Fab is composed of the L chains and N-terminal parts of the H chains, while the fragment Fc is composed of the C-terminal parts of the H chains. The optical rotation patterns of the Fab fragments of the specific and non-specific globulins were different, whereas those of the Fc components were practically identical. The rotatory dispersion data of STEINER and LOWEY (1966) are compiled in Table 17.

The Fc fragments and some of the Bence Jones proteins can be crystallized but the structure has not yet been worked out by X-ray diffraction. Thus the conclusions which could be made on the basis of the Cotton effect data would be highly speculative. A few remarks, however, may be pertinent. It is possible that the fragment Fc has a somewhat different conformation than the fragment Fab. The immunoglobulins and

Table 17. The optical rotatory properties of IgG immunoglobulins and their fragment obtained by digestion with papain. The samples were dissolved in 0,04 M phosphate buffer of pH 7.4 (STEINER and LOWEY, 1966)

Sample	Mol. wt.	b_0 ($\lambda_0 = 212$) degr.	$[m']_{224}$ degr.	$[m']_{231}$ degr.	$[m']_{233}$ degr.
Anti-DNP specific antibody	159,000	+38	−1650	−1340	−1170
Fab fragment of specific antibody	46,000	+64	−2160	−1080	− 770
Fc fragment of specific antibody	50,000	0	− 710	−1570	−1650
Nonspecific IgG	151,000	+24	−1240	−1310	−1260
Fab fragment of nonspecific IgG	51,000	+51	−1510	−1160	−1000
Fc fragment of nonspecific IgG	(50,000)	−13	− 820	−1569	−1650

their fragments seem to be devoid of the α-helix, and the presence of low amounts of the β conformation is likely. IMAHORI (1963) could detect the cross-β structure in oriented films of immunoglobulin by polarized infrared light, but one can object that this structure might have been formed during formation of the film. Troughs at 194 and 230 to 235 nm and peaks at 205 and 210 to 215 nm have also been found in unoriented films of several synthetic polyamino acids, and the presence of the inter-chain antiparallel β and cross-β structures in these polymers has been implied (FAS-MAN and POTTER, 1967). More recently, the pleated-sheet conformation was ascertained in immunoglobulins and amyloid fibrils by TERMINE et al. (1972).

The far ultraviolet Cotton effects of antigen-antibody complexes have also been studied in order to test the possible conformation changes on this interaction. Ross (1967) determined the Cotton effects of diphtheria toxin and its antibody separately as well as the Cotton effects of the complex. Her calculations showed that a conformational change indeed had occurred in this reaction. This change was interpreted as an increase in the amount of β form.

The similarity between the rotatory properties of the urinary Bence Jones proteins and serum γ-globulins was observed as early as 1957/58 [JIRGENSONS, 1958 (1)]. However, comparison of many individual samples showed considerable variation in the Drude constants (λ_c) and specific rotation values at a definite wavelength (JIR-GENSONS, 1959). New impetus in these studies was provided by the discovery of EDELMAN and GALLY (1962) that the Bence Jones proteins are identical with the light chains of the immunoglobulins, and that the individual Bence Jones proteins can be classified into two antigenic types [K and L, or kappa ($\varkappa$) and lambda (λ)] (MANNIK and KUNKEL, 1963; MIGITA and PUTNAM, 1963). This was followed by extensive chemical studies on amino acid sequence and work on conformation. HAMAGUCHI and MIGITA (1964) were the first to report conformational differences between the K and L antigenic types.

During the last decade, several groups of investigators have succeeded in determining the primary structure of human immunoglobulin chains (e.g. PUTNAM et al., 1967; MILSTEIN, FRANGIONI, and PINK, 1967; STEINER and PORTER, 1967; CUNNINGHAM et al., 1968; PONSTINGL, HESS, and HILSCHMANN, 1968). Some of these structures, as regards the apolar nature of side chains and specificity of interaction, have been discussed by KABAT (1968). While the conformational details are unknown, there is no doubt that the geometry of the side (R)-groups along the chain has much to do with the specificity. Significant differences in the amino acid sequences have been found between representatives of the K and L type chains. Even individual samples of the same antigenic type chains differ slightly from each other in primary structure.

More recently, extensive circular dichroism studies on the immunoglobulins and their macromolecular fragments have confirmed and extended the earlier conclusions based on ORD. The findings, to about 1970, have been reviewed by DORRINGTON and TANFORD (1970). Circular dichroism appeared especially valuable in the understanding the effects of the R-group chromophores, as now the CD bands in the near-ultraviolet zone can be identified more reliably than before. IKEDA, HAMAGUCHI, and MIGITA (1968) observed differences between the K and L antigenic type Bence Jones proteins in the 290—310 nm zone, i.e. in the tryptophan chromophores. We observed consistent differences between the K- and L-type light chains in the far ultraviolet, i.e. the K-type proteins yielded a single positive CD band at 200—204 nm, whereas the L-type proteins showed two bands—one at 190—195 nm and another at 197 to 204 nm (KINCAID and JIRGENSONS, 1972). Otherwise the far ultraviolet CD spectra of the Bence Jones proteins are similar to those of the whole 7S immunoglobulins, one case of such spectrum being shown in Fig. 24, as an example of curve resolution.

Interesting species differences have been found by comparing the CD spectra of immunoglobulins from distant species. Fig. 53 shows the CD spectra of individual human versus eel serumglobulins (JIRGENSONS, SPRINGER, and DESAI, 1970). The differences are significant, and it is noteworthy that the human specimens differ from each other much more than the eel globulins. The eel antibody possesses anti-human blood-group H(O) specificity and exhibits interesting specific precipitation with certain monosaccharides. The immunochemical and chemical properties of this protein have been discovered and studied in detail by SPRINGER and DESAI (1971), and the physical properties were described by BEZKOROVAINY, SPRINGER, and DESAI (1971).

The amino acid sequences near the N-terminal parts of the immunoglobulin chains are more variable than at the C-terminals. The peptide links between the variable and constant sequence parts are cleavable by proteolytic enzymes, and the variable and constant halves of Bence Jones proteins have been isolated and studied. Circular dichroism data have indicated some β-structured regions in these fragments and the amount of this order was found to be more in the constant than in the variable halves. The sum of the CD spectra of the isolated constant and variable parts was nearly identical with the CD spectrum of the total protein. These findings suggest that the variable and constant segments of the light chain have more or less independent domains of secondary and tertiary structure (Björk, Karlsson, and Berggård, 1971; Ghose and Jirgensons, 1971), which is in agreement with the ideas of Edelman (1970). According to Litman et al. (1970), the pleated-sheet conformation is mostly within the disulfide loops of the macromolecule. Dorrington and Smith (1972) found significant differences between the CD spectra of immunoglobulin light

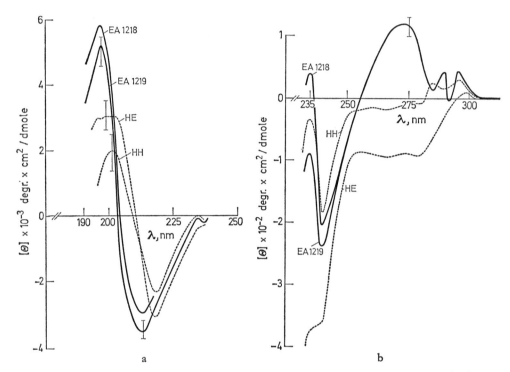

a   b

Fig. 53 a. The ellipticity curves of eel (EA) and human (HE and HH) antibodies in the far u.v. zone. The proteins were dissolved in 0.014 m sodium chloride plus 0.012 m sodium phosphate, pH 7.3. The concentration of the proteins was from 0.0062 to 0.012 per cent; the concentrations were determined by gravimetric microanalysis after correction for moisture and ash. The optical path was 0.10 cm. Each curve represents the arithmetic average of at least three recordings

Fig. 53 b. Circular dichroism of six anti-human blood-group H(O) eel antibodies (EA) and human myeloma 7S globulins HE and HH in the near u.v. zone. The antibodies were dissolved in 0.07 m sodium chloride plus 0.06 m sodium phosphate, pH 7.3. The concentration was from 0.0339 to 0.12%. The optical path was 1.0 cm. Note the different scale expansion on the ordinate as compared to Fig. 53 a. Each curve represents the arithmetic average of at least three recordings. (From Jirgensons, Springer and Desai, 1970)

and heavy chains. They also found that algebraic addition of the subunit spectra did not yield the spectrum of parent immunoglobulin, indicating unique conformational features of the intact protein. According to Abaturov et al. (1969), the infrared spectra of immunoglobulin are in accord with the presence of some β-structured regions in the macromolecule. However, from the magnitude of the CD bands at either 200 or 217 nm, only about 20—30% pleated sheet is indicated (Lee and

JIRGENSONS, 1971). WU and KABAT (1971) have attempted to estimate the possible presence of α-helical segments from known amino acid sequence data, and a low helix content of 12—17% was found permissible as the upper limit of this conformation. Circular dichroism and hydrogen exchange of human γM immunoglobulin has been studied by ASHMAN, KAPLAN, and METZGER (1971).

X-ray diffraction studies on crystallized specimens of immunoglobulins and their fragments have given no conclusive answers regarding conformation because of imperfections in the crystals and absence of helical strands. Thus, on the basis of ORD and CD data, it seems that the major conformation in these proteins is the unspecified aperiodic loop conformation.

More circular dichroism studies are in progress in various laboratories, and clarification of some of the problems involving conformational transitions can be expected. GLAZER and SINGER (1971) found that extrinsic Cotton effects in the CD spectra between 300—500 nm, produced by haptens bound to the active sites of myeloma proteins and antibodies, were characteristic for each protein. This provides a powerful new test for minor conformational differences. Of similar interest are the reversible chemical modifications of the R-groups of immunoglobulins. We observed characteristic changes in the CD spectra of IgG immunoglobulin after mild citraconylation of the ε-amino groups of lysine residues (at pH 7.7 and room temperature and without any denaturing agents). Spectral and hydrodynamic probes indicated partial unfolding and disorganization probably caused by introduction of the $-COO^-$ ions on citraconylation. The modified globulin lost its immunoreactivity. The CD spectrum and immunoreactivity, however, were restored either by increasing the ionic strength or by removal of the citraconyl groups on dialysis at pH 4.0 (NAKAGAWA, CAPETILLO, and JIRGENSONS, 1972). Carbamylation of the ε-amino groups of the lysine residues did not modify the CD spectrum and did not impair immunoprecipitation. This indicates that small chemical changes at the R-groups are not very essential to the immunoreactivity whereas the native conformation is.

7. Effect of Detergents on the Conformation of Nonhelical Proteins

The effect of various anionic detergents on the conformation of nonhelical proteins has been studied extensively by the writer since 1958. The early measurements in the visible and near ultraviolet indicated that the Drude constants (λ_c) of the helical proteins such as serum albumin, were only slightly affected by detergents and that they were enhanced in nonhelical proteins, such as immunoglobulins, soybean trypsin inhibitor, pepsin, and the caseins [JIRGENSONS, 1961 (2)]. It was concluded that treatment of a nonhelical protein with detergent resulted in a *partial transition from the nonperiodic into the α-helical form*. This was ascertained later using other nonhelical proteins by comparing the Moffitt constants (b_0) before and after treatment of the proteins with anionic detergents (JIRGENSONS, 1963).

Studies on the far-ultraviolet Cotton effects confirmed and extended these findings. Some striking instances have already been mentioned: thus Fig. 42 showed the effect of sodium tetradecyl sulfate on β-casein (curve 3). Similar changes, induced by various anionic detergents, were observed also with soybean trypsin inhibitor, chymotrypsinogen, the Bence Jones proteins, immunoglobulins, the histones, and several others. In all cases the change caused by the detergent indicated formation of the α-helix. Moreover, it was found that: (1) the transition is greater the longer the hydrocarbon chain of the detergent, and (2) acid facilitates the helix formation. Some pertinent data are shown in Table 18. Tetradecyl sulfate ion was more effective than dodecyl sulfate, and the latter more effective than decyl sulfate. The effect of acidity was quite conspicuous in all cases. Thus the greatest change was observed with tetradecyl sulfate in acid solutions. Similar results were obtained with many other nonhelical proteins, such as β-casein, the histones, chymotrypsinogen [JIRGENSONS, 1966 (1, 2), 1967 (2)] and the immunoglobulin fragments (LEE and JIRGENSONS, 1971).

Table 18. Effect of decyl (C 10), dodecyl (C 12), and tetradecyl (C 14) sodium sulfates on conformation of soybean trypsin inhibitor. 0.02% aqueous solutions of the protein were treated with the detergents at 50° for 2 hours, and the solutions were cooled to room temperature

Detergent	Concentration, M/l	pH	$[m']$ at 198—200 nm degrees
	0	3.0	$\sim$ 0
C 10	0.016	7.4	+ 5,500
C 10	0.016	2.4	+12,200
C 12	0.005	7.5	+13,600
C 12	0.010	2.3	+18,000
C 14	0.002	6.2	+16,000
C 14	0.004	2.5	+21,500
C 14	0.0016	2.2	+23,000

The mechanism of the protein–detergent interaction can be explained as follows. The negatively charged ions and the hydrophobic "tails" of the detergent molecules both participate in the complex formation between the detergent and protein. The negative ions of the detergent combine with the positive sites of the protein, while the hydrophobic tails are linked to the hydrophobic R-groups of the amino acids by hydrophobic bonding. The hydrophobic tails of the detergent provide a hydrophobic shield around the peptide bonds, thus facilitating the formation and stability of the helices. The promotion of these transitions by acid can be explained as an expansion effect due to electrostatic repulsion of the positively charged sites of the protein, and by the affinity of these sites to the negatively charged detergent ions. The dependence of complex formation and conformational transitions on the length of the hydrocarbon tail confirms the idea *that hydrophobic bonds play a decisive role in protein conformation and that a hydrophobic environment is needed for the formation and stability of the hydrogen-bonded secondary structures.*

The idea of the importance of the ionic interaction is supported by the observations that nonionic detergents are less effective than anionic detergents (JIRGENSONS, 1963). Furthermore, the anionic detergents have only a slight effect on phosvitin which has a high negative charge density.

Many other investigators have studied detergent effects on conformations, *e.g.* EDELHOCH and LIPPOLDT (1960), IMANISHI, MOMOTANI, and ISEMURA (1965), ICHISHIMA and YOSHIDA (1967), LEDERER (1968), and KAY (1970). The importance of the hydrophobic bonding has been emphasized by many, notably KAUZMANN (1959), TANFORD, DE, and TAGGART (1960), and SCHERAGA (1963).

Recent circular dichroism studies have confirmed and extended the rotatory dispersion data on the detergent effects. It has been ascertained that the extent of conformational transition induced by sodium dodecyl sulfate depends on the rigidity and compactness of the protein, and that the effect is enhanced by lowering the pH. While the effect on the backbone of the disulfide-bonded proteins (*e.g.* ribonuclease) it weak, on the unfolded ribonuclease it is spectacular (JIRGENSONS and CAPETILLO, 1970). Two stages can be assumed in the interaction of sodium dodecyl sulfate with a globular protein: 1) binding of the detergent which is accompanied by a partial unfolding, and 2) transformation into a different conformation of relatively hig helix content. Soybean trypsin inhibitor is a typical example of such interaction (Fig. 48). In unfolded proteins such as oxidized ribonuclease the first stage is missing, *i.e.* interaction with the detergent leads directly to folding into ordered conformation. Recent circular dichroism studies on detergent effects in various other proteins have been published by VISSER and BLOUT (1971), DARNALL and BARELA (1971), and ANTHONY and MOSCARELLO (1971). A review on the interaction of detergents and other small molecules with proteins and effects of these small molecules on protein conformation has been provided by PERRIN and HART (1970).

8. Circular Dichroism of Small Polypeptide Hormones

Adenocorticotropin (ACTH), angiotensin, bradykinin, calcitonin, glucagon, secretin, oxytocin, and vasopressin are examples of small polypeptide hormones which are not full-sized proteins. While angiotensin, bradykinin, oxytocin, and vasopressin are relatively very small molecules, ACTH, calcitonin, glucagon, and secretin are composed of 25—32 amino acid residues so that formation of helical folds can be considered and promising results can be expected from the application of chiroptical methods.

Optical activity of the solutions of these medium-sized molecules has been reported by EDELHOCH and LIPPOLDT (1969), GRATZER et al. (1968), and SRERE and BROOKS (1969), among others. According to EDELHOCH and LIPPOLDT (1969), ACTH, glucagon, and the parathyroid hormone solutions yielded CD curves similar to those of disordered proteins, *i.e.* a relatively strong negative CD band was observed at 200 nm and a very weak dichroism at 220—225 nm. SRERE and BROOKS (1969) reported the interesting finding that the CD spectra of glucagon in the far-ultraviolet zone were strongly influenced by the concentration of the polypeptide; while the CD effects were weak and indicative of random conformation in a diluted solution of 0.63 mg/ml glucagon, pronounced negative bands of significant magnitude were observed at a concentration of 12.6 mg/ml at 208—210 nm and at 220—224 nm. Thus the chains of the medium-sized polypeptide hormones can apparently *assume several conformations, depending on the concentration.* A significant amount of the helical structure is likely in the sites where the molecules are accumulated. The formation of the hydrogen-bonded pleated-sheet structure has been indicated even in concentrated solutions of the very small angiotensin (FERMANJIAN, FROMAGEOT, TISTCHENKO, LEICKNAM, and LUTZ, 1972).

Chapter IX

Optical Activity of Structural Proteins

1. Natural Variety of Structural Proteins and Difficulties Involved in Their Study

The rotatory dispersion and conformation of various globular proteins was reviewed in the preceding two chapters. Spectropolarimetry has been applied successfully in the study of these proteins which, as a rule, are soluble. The fibrous structural proteins are more difficult to study by this method, because only a few of them can be dissolved without degradation and possible change of conformation. The main reason for their resistance to mild solvents is their gross conformation, the highly asymmetric fibrous shape of the macromolecules. This can be illustrated by simple geometric consideration. It has been calculated that 8000 carbon atoms in a compact diamond packing form a cube with an edge length of 36.5 angstroms (Å) and a surface area of 7993.5 Å², whereas in a fully extended chain the same 8000 atoms form a 10,200 Å long chain which has a surface area of 177,000 Å². The larger the surface area the greater the number of *contact points* and the possibility of *secondary bonding*. Thus in highly extended polypetide chains, such as in silk or keratin, the secondary hydrogen bonds and van der Waals interactions link the chains at very many segments. Mechanical enmeshing may form structures from which the separation of any of the constituting macromolecules becomes quite difficult. Furthermore, in many protein fibers, such as the keratins, resilin, or elastin, the extended polypeptide chains are linked by covalent bonds. In such cases dissolution without chemical degradation is impossible.

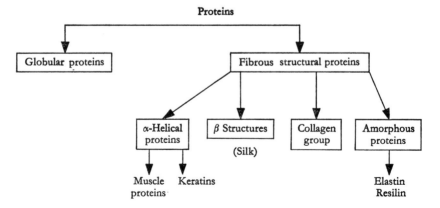

Fig. 54. Classification of the fibrous structural proteins according to conformation

The borderlines between the globular (functional) and fibrous (structural) proteins are rather indefinite and conventional. Thus fibrinogen can be considered as either globular or fibrous protein. While carried around in the blood stream, it is a functional protein; only in the process of blood clotting is it converted into the structural fibrin. The actomyosin of muscle is both a structural and a functional protein.

The conformation of many structural proteins has been studied extensively by X-ray diffraction, and the proteins can be classified on the basis of these data. Four major structural types or classes have been found by this method in unmodified native fibers and tissues: (1) proteins in which the α-*helical* conformation predominates, (2) fibrous proteins possessing the β *structure*, (3) the *collagen group*, and (4) the *amorphous* structural proteins. It is useful to divide the α-helical class into two groups: the soluble muscle proteins and the insoluble keratins. This is illustrated in Fig. 54.

Solutions of muscle proteins and collagens have been studied extensively by CD and rotatory dispersion methods. The keratins and silks are much more difficult objects, because of their insolubility. The strongly crosslinked amorphous elastin and resilin probably are devoid of conformational order. An excellent review on the chemical and physical properties of the structural proteins has been provided by SEIFTER and GALLOP (1966).

2. Structural Proteins with a High α-Helix Content

The fibrous proteins of muscle are the most important in this group. The proteins are extracted from minced muscle tissue usually by means of salt solutions at definite pH's and temperatures, and the extract is fractionated. The *myosins* are the most important of the structural proteins, because they are the major structural components responsible for the muscular function. The myosins are specified by prefixes; there are actomyosins, tropomyosins, paramyosins, and meromyosins, and there has been some confusion in the designation. *Actomyosin*, which is a complex of the fibrous myosin and globular *actin*, was formerly called *myosin*. The fibrous "myosin proper" has been studied by many investigators, but its structure is far from being fully elucidated. Depending on source and method of isolation, the molecular weight of myosin has been found to be between 220,000 and 850,000, and the intrinsic viscosity of the solutions has been reported between 0.5 and 2.5 dl/g. The *meromyosins* are obtained by mild tryptic digestion of myosin, and can be regarded as subunits. There are *heavy (H)* and *light (L) meromyosins*, and it seems that a macromolecule of myosin is composed of one H and two L meromyosins (see the review of SEIFTER and GALLOP, 1966). Then, there are tropomyosins-A and tropomyosins-B. *Tropomyosin-A* now is known to be the same as *paramyosin*, a major structural protein found in the muscle tissue of invertebrates. The molecular weight of paramyosin is 131,000 to 137,00. The high degree of asymmetry of the macromolecules of myosin and paramyosin has been established by a variety of methods including electron microscopy. The α-helical conformation in the rod-shaped macromolecules of the myosins was discovered by COHEN and SZENT-GYORGYI (1957), in the very early period of modern spectropolarimetry. These and later data are compiled in Tables 19 and 20.

According to the data, both the Moffitt constant, b_0, and the magnitudes of the Cotton effect extrema indicate that fraction I of light meromyosin, paramyosin, and

Table 19. Specific rotation and Moffitt constants of muscle proteins

Protein	λ nm	$[\alpha]_\lambda$ degr.	a_0 degr.	b_0 degr.	λ_0 nm	References
H-Meromyosin, rabbit	578	−34.5		−300	210	COHEN, SZENT-GYORGYI (1957)
Do.	D	−30.9	−147	−288	212	McCUBBIN, KAY, OIKAWA (1966)
L-Meromyosin, Fr. I	578	−13.0		−660	210	COHEN, SZENT-GYORGYI (1957)
Do.	D	−15.4	+ 5	−668	212	McCUBBIN, KAY, OIKAWA (1966)
Myosin	D	−28.2	−116	−384	212	Do.
Do.	578	−28.7		−370	210	COHEN, SZENT-GYORGYI (1957)
Paramyosin, *Venus mercenaria*	578	−11.1		−600	210	Do.
Do.			+ 14	−600	212	RIDDIFORD (1966)
Do.			−109	−343	220	Do.
Do. in 7 *M* guanid.-HCl			−476	+ 20	220	Do.
Tropomyosin, rabbit	578	−16.0		−620	210	COHEN, SZENT-GYORGYI (1957)
Tropomyosin, beef cardiac	D	−13.6	+ 9	−623	212	KAY et al. (private communication)

Table 20. Cotton effect data of some muscle proteins

Protein	Solvent	$[m']_{233}$ degrees	$[m']_{199}$ degrees	References
H-Meromyosin, rabbit	Water or 0.10 *M* KCl, pH 7.0	− 7,810	+28,000	McCUBBIN, KAY, and OIKAWA (1966)
L-Meromyosin, Fr. I	0.6 *M* KCl+0.01 *M* phosphate, pH 7.0	−16,200	+71,000	Do.
Myosin, rabbit	0.6 *M* KCl+0.025 *M* Tris, pH 8.0	− 9,650	+42,000	Do.
Paramyosin, *Venus mercenaria*	0.6 *M* KCl, pH 7.2	−15,400	+70,200	RIDDIFORD (1966)
Do.	7 *M* guanid.-HCl+0.6 *M* KCl, pH 7.2	− 2,090		Do.
Tropomyosin, beef cardiac	1.0 *M* KCl+0.067 *M* phosphate, pH 7.4	−15,000	+68,000	KAY et al. (private communication)

the tropomyosins are fully α-helical. The H-meromyosin has less of the helix than the L-meromyosin, and the helix content of the whole myosin nearly corresponds to the sum of the helix of the components.

The stability of the helical secondary structure in the myosins also has been studied by optical rotatory dispersion. Paramyosin appears to be very stable against the denaturing effects of acids and alkalies. According to rotatory dispersion data, no disorganization was detected even in strong acid solutions of pH 1.5 to 2, and only about 15% of the helix was lost in alkaline solutions of pH 12.6. In 5 *M* guanidine hydrochloride a rather complete disorganization could be achieved, but only on heating. At room temperature, some helical order seems to be left even after raising the concentration of the guanidine salt to 7 *M* (RIDDIFORD, 1966). The rod-shaped macromolecule of paramyosin is composed of two helical strands, and it is believed that hydrophobic interactions of the side chains play an important role in the stability of the secondary structure. Also, it is remarkable that the macromolecules of paramyosin, after being disorganized in 7 *M* guanidine salt, reorganize themselves into the helical conformation upon removal of the denaturing agent by dialysis (RIDDIFORD, 1966). This is one of the many examples showing that *the primary structure, i.e. amino acid sequence, determines the conformation.*

Table 21. Circular dichroism of muscle proteins

Protein	$[\Theta]_{222}$ deg·cm²·decimole⁻¹	$[\Theta]_{191}$ deg·cm²·d-mole⁻¹	References
Myosin, rabbit muscle	−23,200	+48,400	OIKAWA, KAY, and McCUBBIN (1968)
H-Meromyosin, rabbit	−18,300	+37,000	OIKAWA, KAY, and McCUBBIN (1968)
Paramyosin, *Venus mercenaria*	−38,900	+87,000	OIKAWA, KAY, and McCUBBIN (1968)
Tropomyosin, rabbit muscle	−37,800	+72,500	STAPRANS and WATANABE (1970)
Troponin, rabbit muscle	−15,200	+25,900	STAPRANS and WATANABE (1970)
Relaxing protein, rabbit	−26,700	+48,400	STAPRANS and WATANABE (1970)

More information has been obtained by circular dichroism measurements (Table 21). The CD curves of all these muscle proteins exhibit a prominent positive band centered at 191 nm and two negative bands at 208—210 nm and 222 nm, respectively. The magnitudes of the bands indicate high α-helix content and confirm the earlier ORD results regarding conformational differences in the various myosins.

Of the other proteins which participate in the muscular activities recent CD studies have been reported on the globular actin (G-actin) by NAGY and STRZELECKA-GOLASZEWSKA (1972). A negative band was observed at 221 nm, and a mean residual molar ellipticity of −8000 was calculated for the band maximum. Similar spectra were observed by MURPHY (1971) wo reported a negative plateau at 218—222 nm with a mean residual molar ellipticity of −11,000. No much change in the far-ultra-violet CD spectrum was observed upon conversion of the G-actin into the polymerized fibrous G-actin; however, significant differences were seen at 250—320 nm between the two forms (MURPHY, 1971). These findings indicate that while there was no significant change in the polypeptide backbone on the conversion, some of the aromatic *R*-groups were dislocated and assumed different positions. These changes probably occurred at the contact sites of the aggregated molecules.

The *keratins* contain all of the common amino acids and an exceptionally large amount of *cystine*. The large number of interchain disulfide bridges in keratin fibers, such as wool, is the main cause of insolubility. If, however, the disulfide bonds are cleaved, a partial dissolution is possible. *Soluble keratin fractions* have been extracted by treating wool with mercaptoacetate and concentrated alkaline urea solutions. The R-SH group-containing fragments are stabilized by blocking the reactive SH groups with iodoacetate, thus obtaining the soluble S-carboxymethylkeratins. The extracted fraction has been separated into several components which differed in composition and physical properties (see J. M. GILLESPIE, 1963). Thus the components which had a high sulfur content differed in conformation from those having a low sulfur content. The macromolecules which had a high sulfur content behaved as random coils, whereas those with low sulfur content seemed to be rod-shaped and of the α-helical conformation. CREWTHER and HARRAP (1967) were able to isolate a subfraction of the low sulfur S-carboxymethylkeratins which had b_0 values of about $-500°$. This was achieved by partial hydrolytic cleavage of the low sulfur component with pronase-P, a protease mixture from *Streptomyces griseus*, and fractionation of the products. The helix-rich fragments, according to CREWTHER and HARRAP (1967), have a particle weight of about 41,000 and an axial ratio of 8 : 1 to 10 : 1. The rod shaped particles seem to be made up of three α-helical chains. In 8 M urea solution the particles were broken down irreversibly to a heterogeneous mixture with molecular weights of about 2000 to 5000. This suggests that many peptide bonds were cleaved in the proteolysis, and that the integrity of the particle in the multistranded structure was secured by secondary bonds. Some analogy between the structural proteins of the keratin and muscle is obvious. The low sulfur helical component in the wool fiber corresponds to the helical L-meromyosin of muscle. In keratin fibers, the helical para-crystalline core of the fiber is surrounded by a matrix of the sulfur-rich amorphous keratin.

3. Optical Activity of Structural Proteins with β Conformation

Since the β structures have been found in many globular proteins (see Chapter VIII), it is likely that they are present also in many fibrous proteins. Thus FRASER and SUZUKI (1965) have reported that parts of the polypeptide chains in feather keratin are highly oriented and are probably in the extended antiparallel β conformation. Natural *silk*, however, is the best known fibrous protein in which the β structures represent the dominant order.

A large variety of silks is known depending on the species of insects which produce them. The common silk fiber from *bombyx mori* has some similarity with the keratin of hair or wool. In both cases the fibers are composed of a highly ordered core and an amorphous matrix. In the instance of silk the amorphous matrix is called sericin; it differs from the ordered core, called *fibroin*, in amino acid composition and conformation. Silk fibroin is composed chiefly of glycine and alanine and is devoid of cystine. The absence of bulky side chains is one of the reasons why the polypeptide chains in the fibroin are so well aligned, so that interchain hydrogen bonds can be formed. The tight packing of the extended chains, which results in relatively strong secondary bonding along the fiber axis, is the chief reason for its mechanical strength and insolubility. The amorphous sericin, however, is much more easily soluble than fibroin, and can be removed by heating with detergent solutions. The remaining fibroin can be dissolved only in a few neutral reagents, *e.g.* in very concentrated, 7 to 9 *M*, lithium bromide.

Iizuka and Yang (1966, 1968) reported their important studies on the optical rotatory dispersion, circular dichroism, and infrared spectra of *bombyx mori* silk fibroin solutions. The fibroin was dissolved in 9.3 *M* LiBr at 37°, and the salt was removed by dialysis. Conformation studies showed that the fibroin in thus prepared aqueous solution had a disordered (random coil) conformation. In the far ultraviolet, the rotatory dispersion curve of the aqueous fibroin had a trough at 205 nm and a peak at 190 nm, while at 198 nm the rotatory power was near zero. However, a profound change in the far-ultraviolet rotatory dispersion occurred upon addition of methanol or dioxane to the aqueous solutions. In a mixed water-methanol (1:1) solvent a partial transition from the disordered into the β conformation took place; the curves had a trough at 229 nm and a positive peak at 205 nm (see Chapters VI to VIII). The corrected mean residue rotation of the fibroin in this solvent was $-2800°$ at 229 nm and $+9900°$ at 205 nm. The magnitude of the positive extremum increased with the methanol concentration, reaching $+15,000°$ in 93% methanol. The Moffitt constant, b_0, of the fibroin in this solvent was $+30$ ($\lambda_0 = 212$).

While the transition from the disordered into the β conformation has been ascertained, many questions remain. It is not known whether the β structures in the methanol solutions were formed between aggregated extended chains (*inter*molecular H bonding) or between the folded segments of unaggregated chains (*intra*molecular H bonding). The tendency of fibroin to form β structures instead of α-helices, however, is quite remarkable. This indicates that the β structure is the most stable order which the polypeptide chains of such particular amino acid composition tend to attain if the competitive interaction with water is reduced by introducing a non-aqueous solvent. In many other proteins, such as β-casein, soybean trypsin inhibitor (see Chapter VIII), or histone (Chapter X), replacement of water with organic solvents leads chiefly to the formation of helical structures. This is true even for globular proteins which have some β conformation in the native state, such as deoxyribonuclease or β-lactoglobulin.

4. Optical Activity and Conformation of Collagen

Collagen, the fibrous protein of connective tissues, is the most abundant protein in animals. It is found in skin and bone, tendons and cartilage, and in most other body structures. As is true of the structural proteins described earlier, there is not just one collagen, but rather many types of collagen differing to some extent in composition and structure. The variety is even greater if the collagenous structural proteins of lower animal species are included for comparison. For example, collagen-like proteins are found in arthropods, in which the major structural material is chitin (J. Gross, 1963).

Collagens differ from all other proteins in amino acid composition and structure of the macromolecules. A peculiar feature in the composition of all collagens is the *high content of proline and hydroxyproline,* amounting to 20% or more. The high content of glycine and alanine in collagens is the only common compositional characteristic that relates them to the silks and keratins. Glycine constitutes about one-third of all amino acid residues and appears periodically in the chain sequences. The collagens are devoid of tryptophan and cystine or cysteine, although traces of the latter have been reported in some collagens of the lower species.

The conformation of collagens has been studied by X-ray diffraction, electron microscopy, infrared spectroscopy, optical rotation, and other methods. Collagen can be extracted from tissues by relatively mild methods, and some of the optical properties have been compared by using a natural collagen-rich fiber and extracted collagen in solution. Also, it has been shown that the higher-order structures observed in collagen fiber can be restored by aggregating the particles of dissolved collagen. The macromolecules are highly asymmetric rigid rods, about 260 nm long and 1.5 nm in diameter. The molecular weight of these units is 300,000 to 350,000, and they have a peculiar structure which is not yet understood in all details. It is, however, established that this fundamental structural unit, sometimes called *tropocollagen*, is present in natural collagen fibers. It is known to be composed of three polypeptide chains wound around each other. The conformational order in each chain is similar (or even identical) to the conformation of polyproline-II as established by X-ray diffraction (see the review of DICKERSON and GEIS, 1969). This order is a left-handed helix having three amino acid residues per turn, *e.g.* glycine-proline-alanine, or glycine-proline-hydroxyproline. This helix differs strongly from the α-helix in dimensions and stability factors. Peptide hydrogen bonds play only a secondary role in the collagen helix, because of the high content of proline and hydroxyproline. Hydrophobic interactions and hydrogen-bonded complexes of water molecules seem to be important in stabilization of this order. Moreover, ester bonds and ionic bonds between the free amino groups of lysine and free carboxyl groups of the dicarboxylic amino acids have been considered. In collagen fibers, the ordered sections are followed by more or less fully disordered segments of the chains. A complete disorganization and separation of the chains occurs at higher temperatures. The denaturation product of collagen is called *gelatin*. Each of the tropocollagen chains, however, is not as stable as in other proteins, because smaller fragments with molecular weights between 17,000 and 20,000 are formed in mild treatments with hydroxylamine under conditions at which the peptide bonds are not cleaved. Thus, bonds other than the ordinary peptide linkages are involved in each of the three tropocollagen chains. An excellent account on these structural studies is provided by SEIFTER and GALLOP (1966), with GALLOP and his associates providing most of the chemical evidence.

Optical rotation and rotatory dispersion of collagen and gelatin solutions have been studied by several groups of investigators, and the early results are reviewed by HARRINGTON and VON HIPPEL (1961). The very early observations with sodium light showed that collagen is an exceptional protein in that it was strongly levorotatory in the native state and that the levorotation diminished on denaturation. The $[\alpha]_D$ values of collagens from various sources were between -280 and $-400°$, whereas those of denatured collagen (gelatin) were -110 to $-135°$. Furthermore, the Drude constant, λ_c, of the collagens was low (204 to 220 nm) and did not change upon conversion of the collagens into gelatin. These early observations indicated that the conformation of collagen is quite different from the conformations found in other structural and globular proteins. Later on, when comparative data on poly-L-proline-II became available, optical rotatory studies helped in ascertaining the presence of the poly-

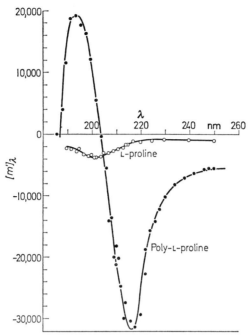

Fig. 55. The far ultraviolet rotatory dispersion of poly-L-proline and proline in aqueous solutions; concentration between 0.015 and 0.147%; optical path 0.1 cm. (From Blout, Carver, and Gross, 1963)

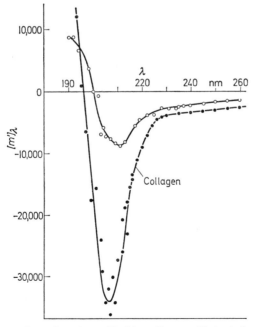

Fig. 56. Rotatory dispersion of native calf skin collagen (filled circles) and gelatin (open circles) obtained by heating the collagen at 50° for 30 min. and cooling. The proteins were dissolved in 0.01 M acetic acid; the concentrations were between 0.0076 and 0.076%. (From Blout, Carver, and Gross, 1963)

proline helix in collagen. HARRINGTON and SELA (1958) found that the $[\alpha]_D$ of poly-proline-II is $-540°$ and that its λ_c is 205 nm. The less negative specific rotation of the collagens, as compared to the polyproline, is consistent with the fact that the polypeptide chains in collagen are only partially organized. From 1950 to 1960, numerous experiments were performed on optical rotation changes that occur on the "melting" of the collagen helices, and that were observed in the reverse process, *i.e.* cooling of the gelatin solutions. A strong negative shift of levorotation was observed on setting (gelation) of warm gelatin solutions, and on long standing of the gels at low temperatures of $+3$ to $+5°$. If, however, a gelatin solution was concentrated by eva-poration at 35 to 40°, or higher, the specific rotation did not change. These obser-vations contributed significantly to the understanding of the structure of gelatin and the gelation process. At sufficiently low temperatures, the polypeptide chains of gelatin tend to aggregate and combine into the tropocollagen-like multichain structures.

The far-ultraviolet rotatory dispersion of poly-L-proline-II and collagen was in-vestigated by BLOUT, CARVER, and GROSS (1963), and the curves are shown in Figs. 55 and 56. Poly-L-proline exhibits a large negative Cotton effect with an inflection point at 203 nm, a minimum at 216 nm, and a positive maximum at 194 nm. The collagen curve was similar, although the position of the minimum was shifted to shorter wave-lengths of about 205 to 210 nm. The magnitude of the negative extremum was diminished strongly on denaturation by heating collagen at 50° for 30 min, as shown by the upper curve in Fig. 56. These findings support the basic principles of collagen conformation, such as the presence of the polyproline type helix and its disorganiza-tion at high temperatures. At the same time, these results imply that the helix in col-

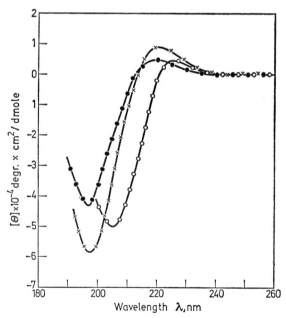

Fig. 57. The $-112°$ C circular dichroism curves for $(Gly\text{-}Pro\text{-}Ala)_n$ ($-\bullet-\bullet-$), poly-L-proline II ($-\bigcirc-\bigcirc-$) and collagen ($-\times-\times-$). Solvent in all cases, ethylene glycol-water (2:1, v/v). (From BROWN, CARVER, and BLOUT, 1969)

lagen is not quite identical with that of the poly-L-proline-II. This is indicated by the shift of the minimum, the difference in crossover (inflection point), and by the extremely high levorotation of the collagen at the trough. If the magnitude of the negative extremum is used for estimation of the amount of this helix, the results would lead to the conclusion that collagen is completely ordered. This, however, contradicts the evidence of the disordered sections in the macromolecules of collagen. The differences in the polyproline and collagen curves are indicative of other transitions in collagen that might be expected on the grounds of different amino acid composition. It is likely that the polyproline helix in the collagen is somewhat distorted or modified. The absence of the α-helix in collagen and gelatin is well demonstrated by these Cotton effects.

Extensive recent circular dichroism studies on collagen and synthetic polyamino acid models for the collagen triple helix have been reported from Professor BLOUT's laboratory. A paper by BROWN, CARVER, and BLOUT (1969), which was briefly mentioned in Chapter VI, reports comparative CD studies on several synthetic poly-amino acids and collagen at low temperatures, and some of the essential findings are shown in Fig. 57. Since the positions of the CD band maxima are the most decisive expressions of conformation, one has to conclude that collagen is much more like (Gly-Pro-Ala)$_n$ in conformation than it is like polyproline. The band maxima of the latter are shifted 5—8 nm to the red in comparison to the positions of the collagen and (Gly-Pro-Ala)$_n$ band maxima. As expected, the optical activity increased at low temperatures, and it was shown that the increase was caused not by temperature-dependent solvent effects but by increasing periodic order as the temperature was lowered. The far-ultraviolet circular dichroism of collagen is consistent with the structure being a triple helix, as suggested for the model (Gly-Pro-Ala)$_n$ polymer in solid state.

Optical Activity of Nucleoproteins and Histones

1. Conformation of Histones

Nucleoproteins are complex macromolecular substances composed of deoxyribo-nucleic acid (DNA) or ribonucleic acid (RNA) and histones. While DNA is known to be the carrier of the genetic code, the role of histones is still imperfectly understood. Two main functions of histones have been considered: (1) regulation of gene activity, either by suppressing DNA replication or by controlling DNA-dependent RNA synthesis, and (2) regulation of the structural integrity of the nucleoprotein complex as a building block for chromosomes. Histones are composed chiefly of alanine, arginine, and lysine; and, because of the excess of basic groups at their side chains (R-groups), they react with nucleic acids. According to ZUBAY and DOTY (1959), the histones in the DNA–protein complex are largely in helical form with their basic side chains electrostatically linked to the phosphate groups of DNA, and they fill the major groove in double-stranded DNA. On the basis of X-ray diffraction studies, WILKINS, ZUBAY, and WILSON (1959) suggested that intra- and inter-DNA histone bridges control the multistranded organization in the chromosomes. Extensive recent reviews on nucleoproteins and histones are available (BONNER and Ts'O, editors, 1964; BUSCH, 1965; PHILLIPS, 1971; HNILICA, 1967, 1972).

The deoxyribonucleoproteins (DNP) or nucleohistones can be obtained from various sources, but those isolated from calf thymus have been investigated in some detail. The histones are liberated from the nucleoprotein complex by treating it with a strong acid. The extracted histones are then purified by dialysis, reprecipitation, further extraction, etc. They are known as heterogeneous mixtures of rather similar molecular species which differ in molecular size and amino acid composition. Only recently have relatively well defined fractions of histones been obtained by using ion exchange chromatography, gel filtration, gel electrophoresis, and selective extraction procedures. Four major fractions of the calf thymus histones have been described: F1, the fraction *very rich in lysine,* and containing 9.4% proline; F2a, the *arginine-rich histone;* F2b, the *moderately lysine-rich histone;* and F3, another *arginine-rich fraction, slow* in starch gel electrophoresis (JOHNS and BUTLER, 1962; HNILICA, 1967). The proline content of the fractions F2a, F2b, and F3 is 3.0, 3.4, and 4.4%, respectively, *i.e.* much lower than in F1. The molecular weight studies have indicated heterogeneity in the otherwise well-characterized fractions. Our earlier rotatory dispersion studies indicated that at low ionic strength, histones are *fully disordered* (JIRGENSONS and HNILICA, 1965), and thus it could be expected that their viscosity behavior would be similar to that of the flexible linear polyelectrolytes. This indeed was ascertained by viscosity measurements, as illustrated in Fig. 58. The dependence of the reduced specific viscosity was found to be linear only in the presence of electrolytes. In the absence of salts, the reduced specific viscosity increased exponentially on dilution. The differences between the viscosity of the various histone fractions also were noteworthy. The fraction F1 yielded much more viscous solutions than the fraction F2b.

The Moffitt constants (b_0) of the various histone fractions were determined by BRADBURY et al. (1965) and by JIRGENSONS and HNILICA (1965). In spite of possible slight differences in the samples used by the British investigators and ourselves, the results agree completely. The b_0 values of histones in aqueous solutions were found to

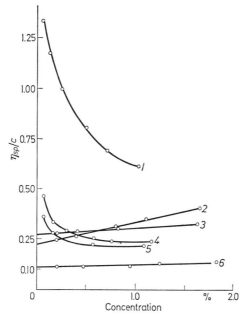

Fig. 58. Dependence of the reduced specific viscosity η_{sp}/c on concentration, c, for various histone fractions in the absence and presence of salts. Curve 1, histone F1 in water, pH 5.7. Curve 2, histone F1 in 0.02 M sodium phosphate. Curve 3, F1 in 0.10 M NaCl. Curve 4, whole, unfractionated calf thymus histone in water. Curve 5, histone F2b in water. Curve 6, histone F2b in 0.10 M NaCl. (From JIRGENSONS, HNILICA, and CAPETILLO, 1966)

be near zero, and negative b_0's of between −70 and −180 were found in 1.0 M sodium chloride solutions. The least negative values were found for the proline-rich, highly viscous fraction F1 and the most highly negative b_0's were found for the fraction F2b. The fractions F2a, F2b, and F3, which contain little proline, were able to form more of the helix than the proline-rich fraction F1. Also, we found that polyvalent anions, such as phosphate, were more effective in organizing the histones than were monovalent anions. Addition of small amounts of DNA (*e.g.* 10% of the weight of the histone) clearly facilitated helix formation. This confirms an early concept of ZUBAY and DOTY (1959) that the histone bound to the nucleic acid is in part in α-helical form. It is possible now to extend and specify this concept as follows: in the presence of anions, the macromolecules of histone fractions F2a, F2b, and F3 in

part assume the α-helical conformation, whereas fraction F1 remains largely disordered. The effectiveness of the ions, especially the phosphates, is due to electrostatic
discharge of the polyvalent histone cations. In the absence of electrolyte, the chains
are stretched out because of electrostatic repulsion. Upon addition of the electrolyte,
the chains can fold up into more compact gross conformation, and some parts of the
chain can assume the α-helical conformation. This is illustrated in Fig. 59, as deduced
from both viscosity and optical rotatory dispersion data.

We extended our measurements into the far ultraviolet, and obtained a wealth of
information. First, in agreement with the results found by the Moffitt method, we
found that the Cotton effect curves of the pure aqueous histone solutions were similar

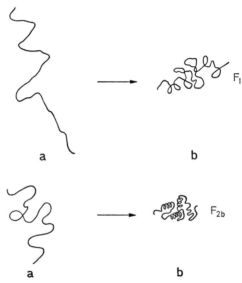

Fig. 59 a and b. Conformational difference between histones F1 and F2b, according to viscosity and rotatory dispersion results. Schematic drawings of the polypeptide chains as they
probably exist in water (a), and transitions into more compact gross conformations (b) under
the influence of salts. The compact coils of histone F1 remain irregular, whereas those of
the histone F2b assume a partial helical conformation. (From JIRGENSONS, HNILICA, and
CAPETILLO, 1966)

to those of the fully disordered polyamino acids and proteins (*e.g.* phosvitin, β-casein,
see Chapter VIII). There was no trough at 232 to 233 nm, but instead a minimum at
206 to 210 nm. Second, upon addition of salt the curves of histones F2a and F2b
took on the form of the curves of partially α-helical proteins, *i.e.* a trough was formed
at 233 nm. In the instance of the fraction F1, however, no trough could be observed
at 233 nm, but instead the minimum at 206 to 210 nm was diminished. Third, on
extension of the measurements still farther into the deep ultraviolet, the pure aqueous
histones exhibited a positive maximum at 191 nm. Fourth, the corrected mean residual
specific rotation of the disordered histones at 233 nm was between −2900° and
−3600°, depending on the composition, *i.e.* it was much more negative than for the
majority of other fully disordered proteins (see Sect. 5 in Chapter VIII). Fifth, the

rotatory power of the histones at 198 to 200 nm was near zero. The latter two findings were of paramount importance for estimation of the α-helix content in those
cases when the helix and disorder are the major conformations. The data are compiled
in Table 22.

The *anionic detergents* were found very active in organizing the histones. The
tetradecyl and dodecyl sulfates could convert even the histone F1 partially into the
α-helical form; but when F1 was compared to F2b, the same amount of detergent

Table 22. The optical rotatory dispersion of histones; pH 5.6 to 6.5
(from JIRGENSONS and HNILICA, 1965)

Protein, solvent	b_0 ($\lambda_0 = 216$) degrees	$[m']_{233}$	$[m']_{220}$	$[m']_{208}$
Histone, unfractionated, H_2O	0	−3500	−3600	− 7,200
Do., 0.10 M NaCl	− 97	−3800	− 500	+ 1,600
Do., 1.0 M NaCl	−146	−4200	+ 800	+ 4,000
Histone F1, aqueous solution	+ 13	−3500	−5400	−10,400
Do., 0.10 M NaCl	− 26	−3500	−4200	− 5,600
Do., 1.0 M NaCl	Non linear	−4100	−2200	− 600
Histone F2a, aqueous solution	0	−2900	−3300	− 7,000
Do., 0.10 M NaCl	−123	−3800	− 300	− 2,300
Do., 1.0 M NaCl, insoluble				
Histone F2b, aqueous solution	0	−3600	−4800	− 8,500
Do., 0.10 M NaCl	−147	−4300	+1100	+ 4,200
Do., 1.0 M NaCl	−172	−4300	+ 400	+ 3,600
Histone F3, aqueous solution	− 8	−3600	−3100	− 900
Do., 0.10 M NaCl	− 88	−3600	− 500	+ 1,500
Do., 1.0 M NaCl	−119	−3500	− 600	+ 2,500

could form more helix in F2b than in F1. This is shown in Fig. 60. Curve *1* shows the
Cotton effects of aqueous histone F1, curve *2* the same with 0.02 M Na-dodecyl sulfate, and curve *3* the comparative effect of the detergent on F2b.

The helix content can be estimated from the magnitudes of the positive extrema,
using the value of 75,000° as standard for 100% α-helical chains (see Chapters VI
and VII). According to Fig. 60, the helix content of histone F1 (0.02% F1 treated
with 0.02 M sodium dodecyl sulfate, pH 6.8) is 18,000/75,000 = 0.24 or 24%, while
for the histone F2b we obtain 26,300/75,000 = 0.35 or 35%. (Since the rotatory power
of disordered histones at 198 to 200 nm is near zero, no other correction is needed.)
Calculation of the helix content of the fraction F2b in 1.0 M NaCl from the Moffitt
constant b_0, according to Table 22, yields about 28% helix. Comparative estimates
of the mole ratios for the various detergents showed that to form 20 to 25% helix in
histone F1 about 5000 moles of decyl sulfate, or 250 moles of dodecyl sulfate, or
50 moles of tetradecyl sulfate, respectively, are needed for one mole F1 (at pH 6.5
to 7.2). Thus the ability to form the helical conformation depends strongly on the
hydrocarbon chain length of the anionic detergent (see p. 122).

The maximum amount of helix which could be formed in any of the histone fractions depended on the *proline content* of the fraction. Thus the maximum amount of

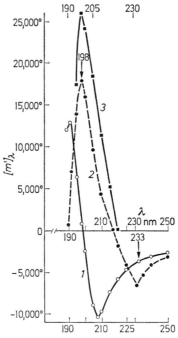

Fig. 60. The corformational transitions of histones effected by sodium dodecyl sulfate. Curve *1*, histone F1, 0.020% in aqueous solution, pH 4.5; optical path 0.10, 0.20, or 0.50 cm. Curve *2*, 0.020% F1 with 0.02 *M* dodecyl sodium sulfate, pH 6.8 (heated for 2 h at 50° and cooled). Curve *3*, histone F2b, 0.02% with 0.02 *M* dodecyl sodium sulfate, pH 7.2 (heated for 2 h at 50° and cooled). [From JIRGENSONS, 1966 (2)]

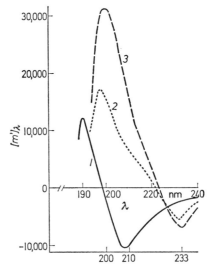

Fig. 61. Effect of n-propanol on the conformation of histone fraction F1. Curve *1*, aqueous solution of F1, pH 4.2. Curve *2*, the histone in 50% n-propanol (*v/v*). Curve *3*, the histone in 98% n-propanol. The concentration of the histone was 0.02 to 0.04%, the optical path 0.050 to 0.20 cm. [From JIRGENSONS, 1967 (2)]

helix which could be formed in the histone F2b (3.4% proline) in acid solutions of sodium tetradecyl sulfate was 42%, whereas the maximum in F1 (9.4% proline) was 31%. An even more complete conversion from the disordered into the helical form could be achieved in organic solvents, but, again, more helix was formed in F2a and F2b than in F1. Estimates based on the Moffitt constants (b_0) showed that in 2-chloroethanol solutions about 67% of the low-proline histone F2b chains are α-helical, whereas in the high-proline F1 the highest helix content was 50%. The conformational transitions of histones in water–alcohol mixtures have been studied by the writer [JIRGENSONS, 1967 (2)]. It was found that in 25% (by volume) n-propanol histone F1 is still largely in disordered state, but a Cotton effect curve typical for partially α-helical protein was obtained of F1 in 50% n-propanol. A trough was observed at 232 nm (magnitude −5800°) and a peak at 197 to 200 nm (magnitude +15,000°). At very high propanol concentrations of 90% and higher, the position of the peak was shifted from 197—200 nm to 199—201 nm, and the magnitude of the extremum increased with rising propanol concentration, while the negative extremum changed but little. This is shown in Fig. 61. Similar results were obtained with methanol and ethanol (JIRGENSONS, unpublished data).

The helix content was calculated only from the peak values, because the estimates from the trough values yielded unreasonably low helix content. This discrepancy is probably due to the very negative correction values for the fully disordered histones as shown in Table 22.

The calculated helix contents, however, should be considered as crude approximations because the shift of the positive extremum from 197 to 200 nm to 199 to 201 nm can be viewed as an indication of the presence of β *structures*. This is supported by the relatively slight changes at the trough as the propanol concentration is increased from 50% to 98% (curves 2 and 3, Fig. 61). Calculation of the helix content from the b_0 values in these cases is more uncertain than calculation of the helix content from the peak values, because straight lines in the Moffitt plots were obtained only in rather narrow spectral zones below 300 nm. Attempts to push up the helix still higher, by adding sodium tetradecyl sulfate, or raising the alcohol concentration to 99%, were unsuccessful. Thus the maximum amount of helix which can be formed is approximately 70% in the low-proline histone F2a and 45% in the high-proline histone F1. The fraction F3 is intermediate in this respect.

The rotatory dispersion results have been confirmed by circular dichroism measurements. In pure aqueous solutions at low ionic strength the CD curves of the various histones have a strong negative band centered at 198—205 nm (JIRGENSONS, unpubl.; TAMBURRO, SCATTURIN, and VIDALI, 1970), and formation of ordered conformation is indicated at higher ionic strength, especially in the presence of long-chain alkyl sulfates or water-miscible organic solvents (JIRGENSONS and CAPETILLO, 1970, and unpublished data). Even more attention has recently been devoted to circular dichroism and conformational transitions upon interaction of histones with nucleic acids, as reviewed in the next section.

2. Optical Activity of Nucleic Acids and Nucleoproteins

The study of the conformation of nucleoproteins is more difficult than the work on proteins because the nucleic acid component too is optically active. As in proteins, the optical activity of the nucleic acids is due to two factors: the asymmetry in each of the many pentose residues and the ordered conformation of the polynucleotide chain.

The first extensive study on the optical rotatory dispersion of nucleic acids was done by SAMEJIMA and YANG (1964). Native salmon DNA displayed Cotton effect curves with three peaks and two troughs as shown in Fig. 62. The peaks were located at 290, 228, and 200 nm respectively, while the troughs were observed at 257 nm and at 215 nm. The magnitude of the specific rotation at the peaks was $[\alpha]_{290} = +1900°$,

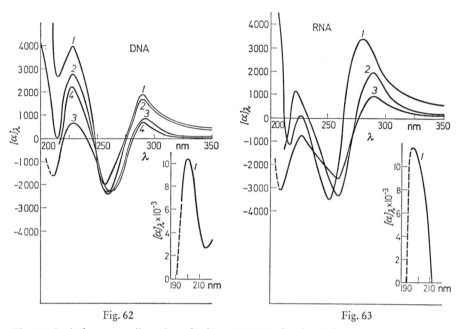

Fig. 62

Fig. 63

Fig. 62. Optical rotatory dispersion of salmon DNA in the ultraviolet region. Curve *1*, DNA in 0.15 *M* NaCl with 0.015 *M* sodium citrate, pH 7 (above 230 nm) or in 0.15 *M* KF, pH 7 (below 240 nm) at 27°. Curve *2*, DNA in citrate-NaCl, boiled for 10 min and cooled. Curve *3*, same as in (*1*) except at 90°. Curve *4*, DNA in 0.15 *M* KF + NaOH, pH 12.3. (From SAMEJIMA and YANG, 1964)

Fig. 63. Rotatory dispersion of rat liver RNA. Curve *1*, RNA in 0.14 *M* NaCl with 0.01 *M* phosphate, pH 7 (above 230 nm) or in 0.14 *M* KF with 0.01 *M* phosphate (below 240 nm) at 27°. Curve *2*, RNA at pH 12.3. Curve *3*, RNA in the same solvent as (*1*) but at 90°. (From SAMEJIMA and YANG, 1964)

$[\alpha]_{228} = +4000°$, and $[\alpha]_{200} = +10,400°$, respectively. The specific rotation at the trough at 257 nm was $-2000°$, while the other minimum at 215 nm had a positive value of $+2600°$. The inflection points (crossover, zero rotation) were observed at 274 and 248 nm, and a probable third inflection could be found by extrapolation at 190 nm. The magnitudes of the extrema were diminished upon denaturation with alkali or heating, an indication that the Cotton effects were caused primarily by the ordered conformation of the polynucleotide backbone. Comparative measurements were made also with calf thymus DNA, and the Cotton effect curves were found to be similar.

In the same series of experiments SAMEJIMA and YANG (1964), compared the far-ultraviolet rotatory dispersion of DNA with that of rat liver RNA. At 250 to 350 nm, the curve of native RNA was similar to that of native DNA. A positive peak was observed at 280 nm, and the $[\alpha]_{280}$ was $+3400°$; at 265 nm the curve crossed the zero line, and a trough of $-3500°$ was observed at 252 nm. The results are shown in Fig. 63. The accuracy below 250 nm was poor because of strong absorbance; a weak Cotton effect seemed to be present near 220 to 225 nm, and a peak appeared at 195 nm. Upon denaturation, both the magnitudes and the positions of the extrema were shifted as indicated in Fig. 63. The positive Cotton effect at 265 to 275 nm is believed to be due to a $n - \pi^*$ transition, and the effects at the shorter wavelengths

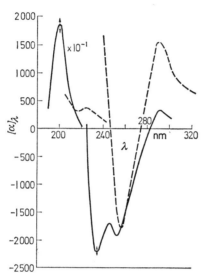

Fig. 64. Rotatory dispersion of calf thymus deoxyribonucleoprotein (DNP) and DNA. The dispersion of DNP is shown by the continuous curve, that of DNA by the dashed curve. The DNP was dissolved in diluted phosphate buffer, pH 7.2, and the DNA was in a diluted phosphate and sodium chloride buffer of pH 7.4. (From ORIEL, 1966)

are related to the $\pi - \pi^*$ transitions. The findings showed that both DNA and RNA possess some ordered secondary structure.

Optical activity of DNA and RNA has been investigated in some detail also in several other laboratories. FASMAN, LINDBLOW, and SEAMAN (1965) studied the Cotton effects of soluble RNA (S-RNA, t-RNA) from yeast, and concluded that this RNA had a highly ordered secondary structure. The main forces responsible for maintaining the structure seemed to be the hydrophobic interactions between stacked base pairs. The asymmetric secondary structure was intact after the amino groups of the bases were blocked with formaldehyde. Thus, hydrogen bonds apparently are not involved in the stabilization. SARIN et al. (1966) compared the far-ultraviolet rotatory dispersion of various highly purified, amino acid-specific t-RNA's, and found significant differences in the Cotton effect curves, e.g. between aspartyl-t-RNA, lysyl-t-RNA and glycyl-t-RNA. Also, they found evidence of conformational change with stabalization in an alternate conformation at temperatures comparable to those in living tissues and cells. This is interesting in that the complete primary structure of some specific t-RNA's is now known (HOLLEY et al., 1965). Simpler polymers, such as polycytidylic and deoxycytidylic acids, have also been studied (FASMAN, LINDBLOW, and GROSSMAN, 1964; ADLER, GROSSMAN, and FASMAN, 1967). The curves were similar to those of natural nucleic acids, but the rotatory power at the peaks and troughs was very high. According to experimental and theoretical studies of McMULLEN, JASKUNAS, and TINOCO (1967), the rotatory dispersion of tobacco mosaic virus RNA indicated that over a wide range of conditions, the rotatory dispersion spectra were found to be a superposition of two basic spectra corresponding to the single-stranded and double-stranded helical forms. Optical rotatory dispersion of nucleotides and nucleosides has also been studied (e.g. BUSH and TINOCO, 1967; MILES, ROBINS, and EYRING, 1967; EMERSON, SWAN, and ULBRICHT, 1967; KLEE and MUDD, 1967).

Optical rotatory dispersion of calf thymus deoxyribonucleoprotein (DNP) has been studied by ORIEL (1966). The DNP of the chromosomes is composed chiefly of DNA and histones; and, in order to assess the histone conformation in the DNP complex, separate series of measurements were made on the DNA isolated from the complex. The curves are shown in Fig. 64. The positive extremum at 290 nm is characteristic for the DNA, as is the trough at 257 nm. The other trough at 233 to 235 nm, is characteristic for the histone component which is partly in the α-helical conformation. The magnitude of this negative extremum is −2140°. To calculate the helix content of the histone, this value must be corrected for the positive rotation caused by DNA in the DNP complex. Knowing the rotatory contribution of the isolated DNA and its content in DNP, ORIEL found −5330° for the histone at 233 nm. He then estimated the helix content of the DNP histone from

$$\% \text{ helix} = 100 \frac{(5330 - 2000)}{14,600} = 23,$$

where 14,600 is the *difference* between the rotatory contribution of the 100% α-helical and fully disordered standard polypeptide, and it is assumed that the corrected mean residual rotation of the fully disordered histone is −2000° at this wavelength. Thus, according to ORIEL, 23% of the histone chains in particulare DNP are in the α-helical form, a value close to the estimates found in salt-containing solutions of histones (see previous section).

More recently, fundamental theoretical studies on the origin of the Cotton effects in nucleic acids and nucleotides have been performed in several institutions, notably by a group led by Professor EYRING (e.g. INSKEEP, MILES, and EYRING, 1970; MILES et al., 1970; CALDWELL and EYRING, 1971). The ORD and CD results on nucleic acids in relation to some theoretical developments have been treated by BRAHMS and BRAHMS (1970). An extensive compilation of the chiroptical data of various nucleic acids is presented in an excellent review of YANG and SAMEJIMA (1969), and comparative CD spectra of the DNA and RNA are shown in Fig. 65.

The heterocyclic bases themselves are optically inactive, but they produce Cotton effects when linked to the asymmetric sugar residues in nucleosides and nucleotides. These effects are conformation-dependent and thus provide a simple and convenient means to study conformational transitions in solution. Also, it is known that in small oligonucleotides the chiroptical effects depend on the sequence. As shown in Fig. 65, the CD spectra of DNA and RNA have a strong CD band centered at 260 nm and corresponding to the major absorption frequency. This CD band depends on conformation and is interpreted in terms of stacking interaction of the nearly planar heterocyclic bases. The fact that organic solvents disorganize the conformation was interpreted as weakening of the hydrophobic and VAN DER WAALS forces among the stacked rings in nonaqueous solvents (YANG and SAMEJIMA, 1969). In all such disorganization reactions by organic solvents, heating, etc., the Cotton effects are diminished and the CD maxima are shifted.

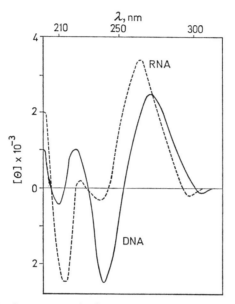

Fig. 65. Circular dichroism spectra of salmon DNA and *Escherichia coli* ribosomal RNA in the far-ultraviolet spectral zone. The aqueous solvents contained 0.005 M Tris, 0.005 M magnesium ions, and 0.1 M KCl. (After YANG and SAMEJIMA, 1969)

GRATZER, HILL, and OWEN (1970) compared the circular dichroism of DNA's from 14 species and attempted to correlate the spectra with the base composition. The CD of the mitochondrial DNA's has also been studied (BERNARDI and TIMASHEFF, 1970). ALLEN et al. (1972) compared the CD spectra of native DNA's with those of synthetic double-stranded polynucleotides and found that the spectra, under favorable conditions, can be used to verify nucleotide sequences.

Extensive recent optical activity studies on nucleohistones and on the interaction of nucleic acids with histones have been conducted in several laboratories. TUAN and BONNER (1969) found that: (1) the conformation of histone in association with DNA approximates to that of free histone in solutions of high ionic strength; (2) the conformation of DNA in natural nucleohistone is different from its conformation when isolated and free; and (3) neither the lysine-rich histone nor the non-histone protein is responsible for the DNA conformation in nucleohistone. SIMPSON and SOBER (1970) found that removal of the lysine-rich fraction from nucleohistone by controlled salt extraction led to a major alteration in the CD spectra, whereas removal of the very lysine-rich or arginine-rich histones had little effect. The CD spectra of the nucleo-histones indicate a significant α-helix content in the bound histones. Extensive CD studies on the histone-DNA complexes have been reported also by Professor FASMAN and his colleagues (*e.g.* FASMAN et al., 1970; ADLER et al., 1971; SHIH and FASMAN, 1971). The CD spectra varied depending on the solvent properties. The conformational change introduced in DNA by complexing with histone F1 was different from the change induced by the other histones. All this work was aimed chiefly at the elucida-tion of the macromolecular structure of chromatin. The CD spectrum of isolated calf thymus DNA was very different from the CD spectrum of DNA in sheared chromatin. MATSUYAMA, TAGASHIRA, and NAGATA (1971) concluded from their CD experiments that the conformation of DNA in chromatin of AH-130 hepatoma cells was sub-stantially the same as that of normal rat liver cells.

3. Conformation of Proteins and Nucleic Acids in Ribosomes, Viruses, and Phages

Optical activity of such complex macromolecular structures as ribosomes and viruses has also been investigated. McPHIE and GRATZER (1966) reported their rota-tory dispersion studies on ribosomes isolated from yeast, *Escherichia coli*, and rabbit reticulocytes, and SARKAR, YANG, and DOTY published a paper (1967) on *Escherichia coli* ribosomes and their constituents. It is known that the basic 30S and 50S particles in the *E. coli* ribosome consist of many different protein molecules associated with the 16S and the 23S RNA components (see SPIRIN and GAVRILOVA, 1969). The results of SARKAR, YANG, and DOTY are shown in Fig. 66. It is obvious that the ultraviolet rotatory dispersion of *E. coli* ribosomes is dominated by the nucleic acid components. The protein component displayed a curve typical for protein of low α-helix content, *i.e.* a trough was observed at 233 nm, and the corrected specific rotation at this point was −4500°. According to McPHIE and GRATZER (1966), the protein in the ribo-somes is partially in the α-helical form, but the isolated protein appeared to be non-helical. These authors determined the amino acid composition of the ribosomal pro-tein and found that with its high alanine and lysine content the ribosomal protein is rather similar to the histones. The proline content of 4.8% also is close to that of the unfractionated calf thymus histone (5.4%). The rotatory dispersion curves of the ribosomes from various sources were significantly different, and the differences seemed to depend chiefly on the protein-to-RNA ratio. McPHIE and GRATZER point out that these differences in the far-ultraviolet rotatory dispersion could be used in gathering taxonomic information on ribosomes.

Extensive studies on the optical activity of viruses and phages have been per-formed by MAESTRE and TINOCO (1967). A large number of different viruses and phages were compared, and significant differences were found in their rotatory disper-sion curves, especially at 220 to 320 nm. As in the case of ribosomes, the main reason for the differences was the protein-to-nucleic acid ratio in the virus. The trough at 233 nm, which is characteristic for a protein with α-helix, was observed only in the viruses which have a large protein coat, *e.g.* the virus f2 and MS2. The α-helix content in the coat proteins was estimated at 5 to 20%. The rotatory dispersion pat-terns of the various viruses, however, depended not only on the protein-to-nucleic acid ratio but also on some structural features, such as mode of packing of the macro-molecules. Hope is expressed that optical activity could be used as a diagnostic technique in the characterization of viruses and phages.

More optical activity and infrared spectral studies on the MS 2 phage were reported by ORIEL and KOENIG (1968). The coat protein subunit spectra and those of

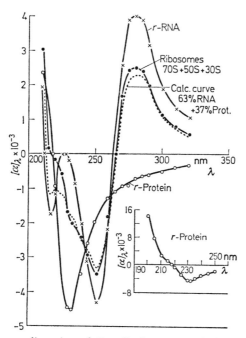

Fig. 66. Optical rotatory dispersion of *E. coli* ribosomes and their constituent RNA and protein components. All measurements were done in the same buffer (0.005 *M* Tris, 0.1 *M* KCl, and 0.004 *M* Mg++). (From SARKAR, YANG, and DOTY, 1967)

the constituent RNA were described upon assembly and heat denaturation, and interesting conformational changes were demonstrated. However, it was admitted that complete interpretation of these observations is very difficult.

Elucidation of the conformational transitions in the complex cellular organelles is considered crucial for an understanding of fundamental processes of life, and much effort has been spent in these endeavors with the aid of chiroptical methods. ADLER, FASMAN, and TAL (1970) examined in detail the circular dichroism of ribosomes and ribosomal RNA; the presence of chelating agents was tested in order to learn about the effect of metal ions; the temperature effects on the conformation also were tested. From the 222-nm CD band amplitude values, it was concluded that the protein in the native ribosomes is 30% helical. Moreover, it was found that heating to 80° had much less effect on the protein than on the conformation of RNA. The subject has been treated also by BRAHMS and BRAHMS (1970).

In the larger and more complex particles, such as mitochondria and ribosomes, the chiroptical effects may be obscured by disturbances caused by light scattering and other effects which will be discussed in Sect. 3 of the next chapter.

Chapter XI

Optical Activity of Glycoproteins and Lipoproteins

1. Glycoproteins and Polysaccharides

Many proteins contain chemically bound carbohydrate. If the carbohydrate component is relatively small, *e.g.* 3 or 4%, its effect on the optical rotatory properties is negligible. Most of such proteins, for instance the immunoglobulins and ovalbumin, have been mentioned previously, and their conformation will not be discussed in this chapter. But there are several proteins which contain 20 to 50%, and more, carbohydrate, and in these cases it is legitimate to question the effect of the carbohydrate component on the optical activity. These carbohydrate-containing proteins are called glycoproteins, mucoproteins, mucoids, and the distinction is made sometimes on the basis of the carbohydrate composition. Since we shall not discuss carbohydrate chemistry, the term *glycoprotein* will be used for all the proteins which contain a substantial quantity of carbohydrate. Hexoses, hexosamines, and sialic acids are the most common building blocks of the carbohydrate components in glycoproteins. Galactose, mannose, and the methylpentose fucose are the most common monosaccharides. Glucosamine and galactosamine appear often, and the amino group of the hexosamine is usually acetylated. The monosaccharides are linked among themselves in oligo- or polysaccharide chains which may be branched. The carbohydrate in the glycoprotein may be either in one polysaccharide moiety or in the form of several such components which have molecular weights of about 500 to 4000. The carbohydrate in glycoproteins is linked to the polypeptide chain by covalent bonds, such as between the anomeric carbon atom of N-acetylhexosamine and the amide of asparagine, or an O-glycosidic linkage involving the reducing group of hexosamine and the hydroxygroup of serine or threonine. A good review on the fundamental structures in glycoproteins is provided by GOTTSCHALK and GRAHAM (1966).

The optical activity of several glycoproteins which have a high carbohydrate content has been studied. The following examples will be reviewed:

1. α_1-acid glycoprotein of serum,
2. fetuin,
3. the haptoglobins,
4. human blood group substances,
5. the ovomucoid of hen egg white,
6. the luteinizing hormone (LH) of sheep pituitary glands,
7. the chondromucoprotein.

To assess the effect of a carbohydrate component in a glycoprotein, the optical activity of pure carbohydrates should be known. RAO and FOSTER (1963) made measurements on several monosaccharides, dextrins, and amylose, and found that in all cases the dispersion effects were weak. According to BEYCHOK and KABAT (1965), simple sugars, oligosaccharides, and amylose displayed simple dispersion curves over the whole spectral region of 589 to 200 nm. The rotatory dispersion followed the one-term Drude equation, and the dispersion constants, λ_c, in all cases were very low, e.g. 147 for maltodecaose, and 153 for amylose. Because the rotatory dispersion of the polysaccharide (amylose) did not differ essentially from that of simple sugars, it was concluded that the *polysaccharide chains do not have a definite fixed secondary structure*. If there is some helical conformation in aqueous solutions, this secondary structure must be loose and changing. However, BEYCHOK and KABAT (1965) found that a Cotton effect with a negative extremum at about 220 nm was displayed by N-acetylated amino sugars and by oligosaccharides and polysaccharides containing such units. The deepest trough was observed with α-L-fucosyl substituted N-acetyl-D-glucosamine, a structure which is present in blood-group glycoproteins. Thus, it is predictable that the *Cotton effects of glycoproteins containing large amounts of N-acetylated hexosamines will be affected by the carbohydrate component*. This has indeed been confirmed by experimental data.

These early studies on the conformation of macromolecular carbohydrates have been extended in width and depth by more recent CD work. The reports up to about 1969 are reviewed by STONE (1969) who has herself made many contributions since 1965 (STONE, 1965; 1971). It appears that the older ORD and CD measurements which end at 200 nm cannot be readily interpreted for structural comparisons, because of the presence of a second CD band centered below 200 nm. The first, negative, CD band (at 205—210 nm) is common to all polysaccharides and amino sugar derivatives, but the characteristics of the second band (at higher frequencies) depend on the fine structure. The polymers containing 4-1 linked amino sugars show a second, positive, band at about 190 nm, whereas those with 3-1 linked amino sugars show a second, negative, band at approximately 188 nm (STONE, 1971). Fig. 67 shows examples of the far-ultraviolet CD spectra of some polysaccharides. The neuraminic and muramic acids (LISTOWSKY, AVIGAD, and ENGLAND, 1970) and some gangliosides also have been studied (STONE, 1971). N-Acetylneuraminic or sialic acid has a relatively strong positive CD band centered at 196 nm and a very weak negative band at about 242 nm; and the residue molar ellipticity of the band at 196 nm is +10,000 deg. cm²/decimole (STONE, 1971). Thus this Cotton effect could partially obscure the peptide band effects in glycoproteins of high sialic acid content.

The rotatory dispersion of α_1-*acid glycoprotein* of serum has been studied by the writer [JIRGENSONS, 1958 (2), 1965] and YAMAGAMI and SCHMID (1967). This glycoprotein (mol. wt. 44,000) contains about 40⁰/o carbohydrate which is present in the form of several polysaccharide chains attached to the polypeptide moiety. According to our measurements (JIRGENSONS, 1965), the rotatory dispersion curve of the α_1-acid glycoprotein had a shallow trough at 228 to 233 nm and a positive extremum at 205 nm. These data indicate that the dominating order of the polypeptide moiety is the β form, and that the α-helix content is negligible. The specific rotation at the trough was −2800°, and at the peak it was +12,000°. YAMAGAMI and SCHMID (1967) investigated the denaturation of the glycoprotein, and found that in 4 M guanidine hydrochloride the negative extremum was shifted from 230 to 220 nm. Also, they determined the rotatory dispersion of the glycopeptide mixture which was split off

the macromolecule and was largely carbohydrate. The dispersion curve of this carbohydrate had a shallow trough at 220 to 225 nm, *i.e.* at nearly the same wavelengths
as in the case of denatured glycoprotein or the N-acetyl hexosamines. The rotatory
power at the trough was only about −900°, and thus it is understandable that this
trough does not show up in the curve for the whole glycoprotein. *The rotatory dispersion thus is strongly dominated by the polypeptide even in the presence of 40%
carbohydrate.* This is further supported by YAMAGAMI and SCHMID (1967) in their
experiments with chloroethanol as solvent: in 50 to 80% chloroethanol−water mixtures the trough in the case of the whole glycoprotein was shifted toward the red and

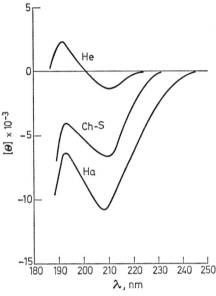

Fig. 67. Far-ultraviolet circular dichroism spectra of polysaccharides. He, heparin; Ch-S,
chondroitin 4-sulfate; Ha, hyaluronic acid. Adapted from STONE (1969, 1971)

deepened, whereas in the instance of the carbohydrate component it was shallower
than in aqueous buffer. According to these authors, the carbohydrate contributes only
about 10% to the rotatory power of the glycoprotein in aqueous buffer solutions, and
even less (5%) in 80% chloroethanol.

Fetuin is a glycoprotein which is found in high concentrations in fetal calf serum.
It contains 22% of carbohydrate, of which 8% is sialic acid. The molecular weight of
fetuin is 48,000. The rotatory dispersion of fetuin has been studied chiefly by KAY and
his associates; thus VERPOORTE and KAY (1966) investigated the Cotton effects of this
glycoprotein betwen 202 to 250 nm. A negative extremum was observed at 228 to
229 nm, and the magnitude of the extremum, in terms of corrected specific mean
residual rotation, was −3100°. This low negative value and the position of the
trough indicate that the helix content in this protein is negligible, and that the
presence of the β conformation is likely. The Drude constant, λ_c, as determined from
measurements in the visible and near-ultraviolet spectral zones, also was low, *i.e.* 228,
which leads to the same conclusion. As the carbohydrate content is lower in fetuin

than in the α_1-acid glycoprotein, we can assume that the weak dispersion and shift of the trough are not caused by the carbohydrate.

Circular dichroism studies of MURRAY, OIKAWA, and KAY (1969) suggest that fetuin has a low helix content in agreement with earlier ORD data. Addition of sodium dodecyl sulfate to the native protein increased the apparent helix content. Neuraminidase-treated fetuin yielded CD spectra similar to those of native fetuin. Similar observations were made by OSHIRO and EYLAR (1969), who also detected very weak CD effects in the 205—240 nm zone and estimated only about 5—10% of helical conformation in native fetuin. Kininogen, a glycoprotein compositionally similar to fetuin, according to KATO et al. (1967) is also practically devoid of helical order.

The haptoglobins of serum are hemoglobin-binding glycoproteins which have molecular weights of 85,000 to 400,000. They contain 20% of carbohydrate, chiefly hexoses, and N-acetylated hexosamines. According to WAKS and ALFSEN (1966), the low values of the Drude constant, λ_c, as well as the low negative Moffitt constants, b_0, are indicative of the absence of significant amounts of the α-helical conformation in these proteins.

The blood-group substances often are called mucopolysaccharides, because the carbohydrate in these macromolecules is the major component. The samples of blood group A, B, and H(0), substances from human ovarian cysts studied by BEYCHOK and KABAT (1965) contained 80% polysaccharide and 20% peptide. Their molecular weight is around 300,000. They all displayed a very weak dispersion, and the curves had a negative extremum at 220 nm. As in the case of the polysaccharide of the earlier mentioned α_1-acid glycoprotein, the rotatory power at 220 nm was weak, *i.e.* between -700 and $-1800°$. The Cotton effect was attributed to the N-acetyl hexosamine content.

A series of blood group-specific glycoproteins and influenza virus receptors from human erythrocyte membranes and secretions have been isolated and well characterized by SPRINGER and his associates (*e.g.* SPRINGER, NAGAI, and TEGTMEYER, 1966; BEZKOROVAINY, SPRINGER, and HOTTA, 1965; SPRINGER, 1967). The circular dichroism of some of these glycoproteins is shown in Fig. 68. Curve CA-825 represents the far-ultraviolet CD of human erythrocyte NN antigen and influenza virus receptor. This glycoprotein contains 44% of peptide, and has a molecular weight of 595,000. The peptide component contains all amino acids, except cystine and tryptophan, and the serine and threonine content is high. The carbohydrate moiety contains large amounts of sialic acid, galactosamine, glucosamine, galactose, and mannose, and small amounts of glucose and methylpentose (fucose). According to Fig. 68, the polypeptide component of this substance has some ordered secondary structure which seems to be chiefly β conformation. The other curve (CA 851) in this illustration shows the CD of a blood group-specific glycoprotein and myxovirus receptor from human meconium (mol. wt. 520,000). Contrary to the sample CA-825, this substance contains only 13% peptide and the optical activity of this substance is very weak. The flat trough at 207 nm is characteristic for the N-acetylated hexosamines of which this substance contains 23.5%.

Fig. 69 shows the far ultraviolet rotatory dispersion of a human blood group MM specific antigen which is also an extremely potent virus receptor. Some α-helix content is indicated by its peak at 200 nm and the trough near 230 nm. However, since the

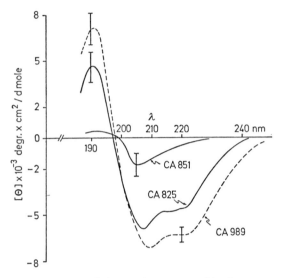

Fig. 68. Circular dichroism curves of the erythrocyte NN blood-group antigen Ca 825, the erythrocyte MM blood-group antigen Ca 989, and the meconium antigen Ca 851. The glycoproteins were dissolved in 0.01 M sodium phosphate buffer, pH 7.2. The mean residue weights used in calculating the molar ellipticities were 177 for Ca 825 and Ca 989 and 226 for Ca 851. The positive CD bands centered at 190 nm may be in part due to the presence of sialic acid.
(From JIRGENSONS and SPRINGER, 1968)

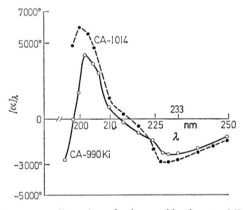

Fig. 69. Far-ultraviolet rotatory dispersion of a human blood group MM specific antigen and virus receptor. Curve CA-1014 shows the rotatory dispersion of the glycoprotein in highly aggregated, very active form, and curve CA-990 Ki represents the dispersion properties of the much less active disaggregation product (subunits) of this substance. (See JIRGENSONS and SPRINGER, 1968; SPRINGER, 1967). The samples were dissolved in 0.01 M phosphate of pH 7.2; the concentration of the glycoproteins was 0.021 to 0.038%

trough extends to 227 nm, some β structure may also be present. The chemical composition of this sample (CA-1014) was similar to the substance CA-825 (Fig. 68), but the molecular weight (1.8×10^6) of this MM antigen was higher than that of CA-825. According to SPRINGER (1967), the relatively large particles of the MM antigen

(CA-1014) are composed of subunits, and the blood-group and anti-influenza virus activities of the disaggregation products are weaker than those of the large aggregate. It is interesting that the rotatory dispersion curve of the disaggregation product, CA-990Ki (mol. wt. 36,000), as shown in Fig. 69, indicates a lesser conformational order than in CA-1014. The difference is small yet conspicuous. The curves are compatible with the presence of a very low but significant helix content and some β conformation in the polypeptide portions of the macromolecules. It is tempting to conclude that the biological activities of these important substances depend not only on the carbohydrate, but also partially on the conformation of the polypeptide.

The conclusions about the very low degree of order which were drawn from the Cotton effect curves are in agreement with results obtained by the Moffitt-Yang method. The Moffitt constants, b_0, were determined from measurements in the near ultraviolet by using a λ_0 of 216. The b_0 values for CA-825, CA-851, CA-1014, and CA-990Ki were -50, 0, -88, and 0, respectively. The b_0 of a blood-group H(0) specific glycoprotein from a pseudomucinous human ovarian cyst (SPRINGER, DESAI, and KOLECKI, 1964) was also determined and found to be $+63$. The dispersion curve of this specimen, which contains a relatively small peptide component, had a shallow trough at 220 to 230 nm, and the specific rotation at 225 nm was $-2040°$.

The *ovomucoid* (mol. wt. 27,000) from hen egg white has been investigated by the writer (unpublished data) and by TOMIMATSU and GAFFIELD (1965). This glycoprotein contains approximately 23% carbohydrate, and the polysaccharide portions are rich in N-acetyl-D-glucosamine. In our early work the rotatory dispersion was studied in the visible and near ultraviolet spectral zones and a Drude constant $\lambda_c = 228$ was found. This value did not change in acid solutions, even at pH as low as 1.1, but the λ_c decreased to 220 in alkaline solution of pH 12.0. TOMIMATSU and GAFFIELD (1965) worked with different commercial samples of ovomucoid. They found a λ_c of 231, a b_0 of -86 (with $\lambda_0 = 212$); and a trough at 230 nm and a peak at 198 nm in the far ultraviolet. From these data they concluded that ovomucoid probably contains about 20 of the α-helix. Our measurements in the far ultraviolet indicate a much lower amount of this conformation. A trough was observed in the curves at 230 nm which is in agreement with TOMIMATSU and GAFFIELD, but the specific rotation at this wavelength was only $-4200°$. Further more, the peak was observed not at 198 but at 200 to 201 nm, and the specific rotation at that point was $+7500°$. The Moffitt constant, b_0, which was determined from measurements between 325 and 245 nm ($\lambda_0 = 216$), was found to be zero. From these data, it is more likely that portions of the polypeptide chain in ovomucoid have the β structure and that the α-helix content is negligible or very low. The reason for these discrepancies is not known.

The *luteinizing hormone*, LH, also called interstitial cell-stimulating hormone, is a very interesting glycoprotein. Most LH preparations from sheep pituitary glands have been studied, notably by WARD, WALBORG, and ADAMS-MAYNE (1961), WARD et al. (1967), and by PAPKOFF et al. (1965). The molecular weigth of this glycoprotein is 28,000 to 32,000, and in acid solutions it dissociates into unequal subunits (LI and STARMAN, 1964). Quantitative analysis of the carbohydrates yielded 8 residues of N-acetylglucosamine, 3 residues of N-acetylgalactosamine, 7 residues of mannose, 2 of galactose, and 1 to 2 of fucose (WALBORG and WARD, 1963). In the polypeptide component, the high contents of proline (25 residues per mole) and cystine (8 to 10 residues per mole) are noteworthy. The serine and threonine contents are also high.

Our early measurements of rotatory dispersion in the visible and near-ultraviolet spectrum yielded Drude constants, λ_c, of 213 to 218, indicating the absence of α-helix in this protein (JIRGENSONS, 1960). Neither the specific rotation nor the dispersion constant was altered if the hormone was denatured by $7\,M$ urea or $4\,M$ guanidine thiocyanate. Later work on the far ultraviolet Cotton effects fully substantiated the earlier conclusions. The curve had a shallow trough at 220 to 228 nm, and the rotatory power at 200 nm was near zero (JIRGENSONS, 1964).

Circular dichroism of ovine luteinizing hormone and its subunits has been studied in at least three laboratories (JIRGENSONS and WARD, 1970; PERNOLLET and GARNIER, 1971; BEWLEY, SAIRAM, and LI, 1972). The far ultraviolet CD of the hormone is somewhat similar to the earlier described spectra of the trypsin inhibitors from soya bean and lima bean (Chapter VIII, Sect. 4). In the CD spectrum of LH and its subunits, there is no negative peak or even a shoulder in the curves at 222 nm, and no positive band was observed at 190—195 nm. Thus CD confirms the nonhelical nature of the luteinizing hormone; also, the CD data exclude a large amount of the pleated-sheet structure. Interesting differences in the subunit conformation are expressed in the 230—330 nm zone. This is expected, because of significant differences now known to exist in the composition and primary structure of the subunits (PAPKOFF, SAIRAM, and LI, 1971; LIU et al., 1972 a, b). Although the conformational details are unknown, the relatively low intrinsic viscosity of LH and its subunits (0.044—0.046, WARD et al., 1967) indicates a rather compact conformation. According to recent studies on the positions of the disulfide bridges and sequence, Professor WARD has developed a molecular model of the hormone in which the polypeptide backbone is mostly in the form of several loops with only very limited hydrogen bond formation (personal communication).

The conformation of *chondromucoprotein* from bovine nasal septum has been studied recently by EYRING and YANG (1968). Chondromucoprotein is a protein-polysaccharide complex of molecular weight 1.8×10^6 composed of about 20% protein and 80% polysaccharide. The complex behaved as a highly asymmetric polyelectrolyte in pure aqueous solution, *i.e.* the reduced specific viscosity increased on dilution, and the expanded particles contracted in the presence of salts. Optical rotatory dispersion and circular dichroism measurements indicated the absence of any helical or other secondary structure. The macromolecules behaved as random coils.

In summary: it is remarkable that all of the *glycoproteins* thus far investigated *are either devoid of the α-helical conformation or have only little of it. The β structures are not dominating either. The presence of the carbohydrate*, which is distributed in several places in the macromolecule, may be the chief reason for the low content of the hydrogen bonded secondary structures. *The carbohydrate makes the whole structure more hydrophilic than it is in most of the ordinary proteins. Thus, the interior of the glycoprotein molecule is strongly exposed to water, so that the shielding provided by the hydrophobic side chains of amino acids is not sufficient to protect the hydrogen bonds of the possible helices from competitive interactions with water.*

2. Lipoproteins

The lipoproteins contain a substantial amount of lipid, mostly in the form of a phosphatide, cholesterol and its esters, and triglyceride. Since the lipids can be removed from the lipoprotein by extraction with mild solvents, such as mixtures of alcohol and ether, the dispersed particles of the lipoproteins should not be considered as macromolecules. Thus the "molecular weights" of the lipoproteins are in fact micellar weights; the particles of lipoproteins, *e.g.* in blood, are complex aggregates of one or more protein components with all sorts of micellar lipids.

Lipoproteins are classified according to their density, which is relatively low, so that in the process of ultracentrifugation in buffer solutions the lipoproteins do not sediment but float on top. The *very low-density* lipoproteins have density values of 1.006 to 1.019, and they are composed largely of lipids covered by a thin layer of protein. These particles have the electrophoretic mobility of the α_2 globulins. The relative weight of these lipoproteins in solution is over 5 million. The *low-density* β_1 lipoproteins have in aqueous solutions somewhat smaller particles of density between 1.019 and 1.063. These are also composed largely of lipid. The *high-density lipoproteins* contain 30 to 58% of protein, they behave in an electrical field as α_1 globulins, and have densities of 1.063 to 1.21. They can be fractionated further into high density and very high density components. The former have the "molecular weight" of 435,000 and the latter of 195,000. Concise compilations of other pertinent data on these substances can be found in the book of SCHULZE and HEREMANS (1966) and in the article of GIDEZ (1967).

The optical rotatory dispersion of *high-density lipoproteins* from human serum has been studied by SCANU (1965). The rotatory power was measured in the visible, near-ultraviolet and far-ultraviolet spectrum and the measurements indicated a high α-helix content by all criteria. Thus, the Moffitt constant (b_0) of the very high density lipoprotein HDL$_3$ was -516 (determined with a $\lambda_0 = 212$), the corrected mean residual specific rotation at 233 nm was $-9100°$, and the $[m']_{199}$ value was $+56,000°$. If the α-helix content is calculated from the latter, one gets about 75% of α-helix. Removal of the lipid resulted in slight disorganization, as judged from the changes in optical activity and viscosity. It is remarkable that the lipoproteins were relatively stable toward the disorganizing effects of 8 M guanidine hydrochloride, and that the changes produced by the denaturing agent were reversible. Thus, the b_0 of the high density lipoprotein HDL$_2$ in 8 M guanidine-HCl was -200, and it returned to -507 upon removal of the guanidine salt by dialysis. If the lipid was removed before denaturation, the disorganization was complete ($b_0 = 0$, and $[m']_{233} = -2200°$). However, even in this experiment the protein returned to its native conformation when liberated from the guanidine salt; the b_0 of the recovered protein was -501, and the $[m']_{233}$ value was $-7500°$.

These results are quite remarkable. First, we have here another example which shows that the primary structure determines the conformation. It can be safely assumed that in 8 M guanidine–HCl the polypeptide chains were fully disordered (the intrinsic viscosity was 0.38 dl/g); yet they assumed the α-helical conformation after removal of the guanidine to a high degree. Second, the *very high α-helix content in the lipoprotein* is noteworthy. It is tempting to conclude that the presence of the lipid is one of the chief reasons for this high degree of helicity. It is known that in the high density lipoproteins, the lipid does not form a compact core but rather is distributed between the subunits of the protein. Some of the polar groups of the lipid may even be on the surface of the complex micelle, while the nonpolar groups form hydrophobic links with the nonpolar residues of the peptide. The hydrogen bonds of the *protein helices thus are protected by the lipid from competitive interactions with water.* This is a situation diametrically different from that in glycoproteins. The optical activity contribution of the nonprotein component in the lipoproteins is slight.

The very high α-helix content in the high-density α-lipoproteins was confirmed by CD studies of SCANU and HIRZ (1968). Later, SCANU (1969) found that heating from 20° to 80° diminished the helix content of the lipoprotein only slightly, whereas the same heat treatment of the delipidated protein resulted in more complete denaturation. Moreover, it was observed that sodium dodecyl sulfate at high concentrations

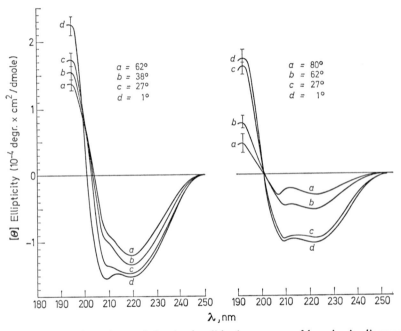

Fig. 70. Temperature dependence of the circular dichroism spectra of low-density lipoproteins from human serum. The curve system on the *left* represents the mean residue ellipticities of the native lipoprotein, whereas the curves on the right show the behavior of the delipidated protein component (apoprotein). (From SCANU et al., 1969)

had the same stabilizing effect on the proteins as the lipid. This satisfactorily confirms the concept that the high helix content is due to the protective shielding of the hydrogen bonds by hydrophobic groups of either the lipid or detergent.

The low-density β-lipoproteins were also studied by circular dichroism and other optical methods (GOTTO, LEVY, and FREDERICKSON, 1968; SCANU et al., 1969; BROWN et al., 1970). More inhomogeneity was found in the protein of these lipoproteins than in the heavy ones. The various chromatographically separated protein fractions had either high or low helix content and a significant portion of the polypeptide backbone had the pleated-sheet structure. As in the case of the high-density lipoproteins, the apoproteins from the low-density lipoproteins were more sensitive to heat denaturation than the whole, lipid-containing proteins (SCANU et al., 1969). The CD spectra are shown in Fig. 70.

While in the earlier optical activity studies of the heavy lipoproteins attention was directed mostly to the far-ultraviolet zone, detailed CD experiments in the 250—320 nm zone have been reported recently by LUX et al. (1972). These near-ultraviolet spectra showed peaks at 258, 264, 283.5, and 290.5 nm. Delipidation reduced and markedly altered the ellipticity bands. The bands at 283.5 and 290.5 nm,

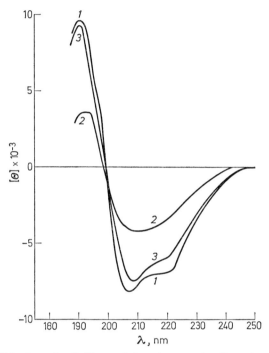

Fig. 71. Circular dichroism of a lipid-containing glycoprotein (lipopolysaccharide endotoxin receptor). Curve *1*, native lipo-glycoprotein in aqueous solution. Curve *2*, delipidated glyco-protein in aqueous solution, pH 6.3. Curve *3*, 0.0068% delipidated material in 0.005 *M* sodium dodecyl sulfate, pH 6.8. The residue molar ellipticities were calculated taking a mean residue weight of 145 for the glycoprotein (SPRINGER et al., unpubl.)

tentatively assigned to one or more tryptophan chromophores, where shifted to 286 and 292 nm, respectively, and the latter peak changed its sign. The near-ultraviolet CD spectra of the native lipoprotein were partially restored by recombining the protein component with phosphatidylcholine and they were almost completely restored by recombination with both phosphatidylcholine and cholesteryl oleate. The near-ultraviolet CD spectrum was more sensitive to these lipid-induced changes than the spectrum in the far ultraviolet. From the latter, LUX et al. (1972) estimated a helix content of 70% and 5—15% of pleated-sheet conformation; delipidation decreased the helix content to about 50%.

Regarding the helix-promoting effect of lipid *versus* the anti-helix effect of carbohydrate component in proteins, it was of interest to test a lipid-containing glycoprotein. SPRINGER et al. (in press) have isolated such an object from human erythrocytes. The complex is composed of about 60% protein, 30% carbohydrate, and 10% lipid; its molecular weight is 228,000 and it consists of subunits. The far-ultraviolet circular dichroism spectrum is shown in Fig. 71. As expected, the spectrum indicates a moderately ordered conformation. The disorganizing effect of the hydrophilic carbohydrate (about half of which is sialic acid) is largely compensated by the protective effect of the lipid. A significant amount of the α-helical and β-conformations is indicated by the sloping plateau at 215—225 nm. The positive band at 190—195 nm may be partially caused also by the sialic acid content. Removal of the lipid resulted in some disorganization (curve 2, Fig. 71). When sodium dodecyl sulfate was added to the delipidated protein, the original native conformation was recovered almost completely (curve 3, Fig. 71).

3. Membranes and Other Particulate Suspensions

Cell membranes are more complex than lipoproteins, yet membrane dispersions too have been studied by chiroptical methods. If properly liberated from extraneous inclusions and prepared in the form of colloidal dispersions of low turbidity, they represent fairly reproducible systems. The membranes are composed chiefly of proteins and lipids, but carbohydrates and other minor components must also be considered.

Several papers on the optical activity of membrane suspensions were published in 1965—1968, but the results were difficult to interpret because of unusual location of the Cotton effects. The negative CD band of the helix located at 222 nm was shifted to about 225 nm and the positive band at 190—200 nm was weak (URRY, MEDNIEKS, and BEJNAROWITZ, 1967). These distortions were explained in 1968—1971 when URRY and JI (1968) considered the anomalous absorption and light-scattering effects caused by the particles. The so-called absorption flattening results in a decrease of absorbance due to decrease in the effective concentration of the absorber; scattering also leads to loss of light intensity. It also has been suggested that the particles can scatter the left-circularly polarized light to a somewhat different degree from the rigth-circularly polarized light (JI and URRY, 1969; URRY, 1970). All this causes the above-mentioned distortions, as has been confirmed by experiments with the same particulate substance disintegrated by sonication or other methods (SCHNEIDER, SCHNEIDER, and ROSENHECK, 1970; URRY, MASOTTI, and KRIVACIC, 1971; GLASER and SINGER, 1971 b). While the turbid suspensions yielded distorted CD spectra in the far-ultraviolet zone, the reduction of turbidity led to the normal CD curves having the bands characteristic for the α-helix located at the normal frequencies, *e.g.* the "anomalous" 225 nm band was shifted back to 222 nm. Red-cell membranes and mitochondria have been the chief objects of study, and the circular dichroism spectra, corrected for the described anomalies have indicated about 40—50% of α-helix in these highly organized particles. Suspensions of the sarcotubular vesicles from rabbit muscle (MASOTTI et al., 1972), however showed lower ellipticities and the changes in the spectra after various treatments indicated that they have a different structural organization from red-cell membranes. Tubulin, a microtubule protein from pig brain, according to the CD spectra, had only 22% of α-helix and 30% of pleated-sheet conformation (VENTILLA, CANTOR, and SHELANSKI, 1972). Chloroplasts, for example the thylakoid membrane dispersions (from purple bacterium *Rhodopseudomonas spheroides* or the blue-green alga *Oscillatoria chalybea*), displayed CD spectra with characteristics similar to those of other particulate systems (MENKE, 1970). A significant amount of the α-helix was found in these particles and interesting conformational transitions were observed upon various treatments of these particulate systems.

Chapter XII

Concluding Remarks

Summarizing briefly the major conclusions, the following statements can be made.

During the last ten years, optical rotatory dispersion (ORD) and circular dichroism (CD) have established themselves fully as leading methods in the study of conformational transitions of peptides, proteins, and various other biopolymers in solution. The X-ray diffraction analysis of solvent-containing protein crystals has confirmed on at least 20 proteins that the conclusions on conformation derived earlier from ORD and CD were essentially correct. (The prospects for clarifying the few remaining disagreements are good.) In contrast to earlier views, a variety of structural orders has been disclosed in the macromolecules of proteins. Ten years ago the α-helix and random coil were considered to be the major conformations, but now a large variety of helices have been disclosed by X-ray structural analysis. Also, it is established that a perfect helical strand is the exception rather than the rule in globular proteins. The pleated-sheet conformation now appears more prominent than was previously assumed, and both parallel and antiparallel β-forms have been found. Most important, however, is the discovery that the major conformational order in several rigid globular proteins is neither helix nor pleated sheet but some aperiodic order. These loop-shaped segments are specific for each protein, and they are stabilized chiefly by hydrophobic forces. Each protein has a unique macromolecular geometry, as is expressed in the CD curves.

One of the most important developments in the last five years is the realization that the nonperiodic folding of the polypeptide backbone chain should not be equated with the statistical random coil or disordered conformation. The latter conformation is continuously changing, whereas the nonperiodic order in globular proteins is rigid and unique for each protein. Also it was discovered that the CD curves of the presumably random-coil polyamino acids which carry an electrical charge at each residue differ from the CD curves of fully disordered proteins. Because of this, and the discovery of the aperiodic loop conformation, the polyamino acids appear unsuitable as protein models, especially for nonhelical proteins. This is supported also by the variety of chiroptical effects displayed by pleated sheets of various length and width.

The difficulties involved in the theoretical efforts made to solve these problems have been well summarized in a recent publication of MADISON and SCHELLMAN (1972). The theory is able to predict the CD curves of α-helical strands and pleated sheets. However, the α and β structures in proteins differ from the models in being very short and distorted. It is concluded that the 222 nm band of the α-helix is the most adequate measure for the α-helix content. The circular dichroism of the β-regions of proteins differs markedly from the ideal pleated sheet because of a breakdown in symmetry; thus, polyamino acids are not good models for the estimation of this conformation. The theory has even greater difficulties with the polyproline II helices and, of course, with the aperiodic loops. There are considerable complexities too in the statistical calculations to treat random-coil effects, *i.e.* to predict the CD curves (*e.g.* ZUBKOV et al., 1970). Progress in solving the theoretical problems associated with random coils and aperiodic conformations is likely to be slow. The same is true of the possibility of predicting the conformation of macromolecules of proteins from known amino acid sequences.

Although chiroptical methods alone are unable to give all the answers on structural details, the methods have yielded a wealth of information on conformational transitions in solution (*e.g.* effects of pH, ionic strength, temperature, various additives, etc.). Once the significance of the major Cotton effects is established, important conclusions can be drawn from the observed changes in the spectra. The ease of the experimental procedure and increase of sensitivity of the instruments has greatly facilitated the spreading of the methods in biochemistry and biology. The study of the circular dichroism in the near-ultraviolet zone has made exceptional progress. Studies on the fine structure of CD spectra at low temperatures, paralleled by similar determinations of absorption, have contributed heavily to the identification of the bands, *i.e.* to correlation with the particular chromophore. However, much more remains to be done. The band overlap of the many chromophores, each usually in a different microenvironment in the macromolecule, poses often very difficult problems.

As the data accumulate, some important relationships are now becoming more obvious, *e.g.* effects of carbohydrate components on the CD spectra and conformation of a glycoprotein, effects of a lipid in a lipoprotein, turbidity effects, interactions of the various histone fractions with DNA, etc. Although many problems in the interpretation of the CD spectra are still unsolved, the prospect of overcoming the difficulties looks encouraging. Many problems can be solved with the aid of the equipment and principles now available, but even greater progress can be expected from further improvements in instrumentation and theory, and the correlative analysis of the experimental material already available and still to be provided.

References

ABATUROV, L. V., NEZLIN, R. S., VENGEROVA, T. I., VARSHAVSKY, J. M.: Conformational studies of immunoglobulin G and its subunits by the methods of hydrogen-deuterium exchange and infrared spectroscopy. Biochim. et Biophys. Acta **194**, 386—396 (1969).

ADAMS, M. J., FORD, G. C., KOEKOEK, R., LENZ, P. L., McPHERSON, A., ROSSMANN, M. G., SMILEY, I. E., SCHEWITZ, R. W., WONACOTT, A. J.: Structure of lactate dehydrogenase at 2.8 A resolution. Nature **227**, 1098—1103 (1970).

ADKINS, B. J., YANG, J. T.: Difference spectropolarimetry as a probe for small conformational changes. Biochemistry **7**, 266—271 (1968).

ADLER, A., GROSSMAN, L., FASMAN, G. D.: Single stranded oligomers and polymers of cytidylic and 2'-deoxycytidylic acids. Comparative optical rotation studies. Proc. Nat. Acad. Sci. US **57**, 423—430 (1967).

ADLER, A. J., FASMAN, G. D., TAL, M.: Circular dichroism of *Escherichia coli* ribosomes: effect of heating and metal ions. Biochim. et Biophys. Acta **213**, 424—436 (1970).

ADLER, A. J., SCHAFFHAUSEN, B., LANGAN, T. A., FASMAN, G. D.: Altered conformational effects of phosphorylated lysine-rich histone (fl) in fl-DNA complexes. Circular dichroism and immunological studies. Biochemistry **10**, 909—913 (1971).

ADLER, A. J., FASMAN, G. D.: Optical rotatory dispersion as a means of determining nucleic acid conformation. In: Methods in enzymology (eds. L. GROSMAN, K. MOLDAVE), vol. XII (B), pp. 268—302. New York-London: Academic Press 1968.

ADLER, A. J., GROSSMAN, L., FASMAN, G. D.: Circular dichroism of cytosine dinucleotide monophosphates containing arabinose, ribose, and deoxyribose. Biochemistry **7**, 3836—3843 (1968).

ADLER, A. J., HOWING, R., POTTER, J., WELLS, M., FASMAN, G. D.: Circular dichroism of polypeptides. Poly(hydroxyethyl-L-glutamine) compared to poly(L-glutamic acid). J. Am. Chem. Soc. **90**, 4736—4738 (1968).

AKI, K., TAKAGI, T., ISEMURA, T., YAMANO, T.: Optical rotatory dispersion and circular dichroism of D-amino acid oxidase. Biochim. et Biophys. Acta **122**, 193—201 (1966).

D'ALBIS, A.: Effets Cotton de la trypsine bovine. Compt. rend. Acad. Sci., Paris **262**, 1888 a 1891 (1966).

ALLEN, F. S., GRAY, D. M., ROBERTS, G. P., TINOCO, I., JR.: The ultraviolet circular dichroism of some natural DNAs and an analysis of the spectra for sequence information. Biopolymers **11**, 853—879 (1972).

ALMQUIST, H. J., GREENBERG, D. M.: The influence of pH on the optical rotation of proteins. J. Biol. Chem. **105**, 519—522 (1934).

ANFINSEN, C. B.: The tertiary structure of ribonuclease. In: Brookhaven Symp. in Biology, vol. 15 (Enzyme models and enzyme structure), pp. 184—198. Upton, New York: Biology Dept., Brookhaven Natl. Lab. 1962.

ANTHONY, J. S., MOSCARELLO, M. A.: A conformation change induced in the basic encephalitogen by lipids. Biochim. et Biophys. Acta **243**, 429—433 (1971).

APPLEBURY, M. L., COLEMAN, J. E.: *Escherichia coli* alkaline phosphatase. Metal binding, protein conformation, and quaternary structure. J. Biol. Chem. **244**, 308—318 (1969).

APPLEQUIST, J. B., DOTY, P.: α-Helix formation in poly-ε-carbobenzoxy-L-lysine and poly-L-lysine. In: Polyamino acids, polypeptides, and proteins (ed. M. A. STAHMANN), pp. 161—176. Madison, Wisconsin: University of Wisconsin Press 1961.

ASHMAN, R. F., KAPLAN, A. P., METZGER, H.: A search for conformational change on ligand binding in a human γM macroglobulin-I. Circular dichroism and hydrogen exchange. Immunochem. **8**, 627—641 (1971).

162   References

Askonas, B., Butler, J. A. V., Jacob, F., Phillips, D. C., Sachs, L., Sherbet, G. V.: Biological systems at molecular level. Report, Commission of Molecular Biophysics, Internatl. Union of Pure and Appl. Biophysics, Naples, Italy, Sept. 8—11, 1965. Nature **208**, 1048 to 1050 (1965).
Astbury, W. T.: Fundamentals of fibre structure. Oxford: University of Oxford Press 1933.

Balasubramanian, D., Wetlaufer, D. B.: Optical rotatory properties of diketopiperazines. J. Am. Chem. Soc. **88**, 3449—3450 (1966).
Barth, G., Voelter, W., Bunnenberg, E., Djerassi, C.: Magnetic circular dichroism studies. XVII. MCD spectra of proteins. A new method for quantitative determination of tryptophan. J. Am. Chem. Soc. **94**, 1293—1298 (1972).
Bayley, P. M., Radda, G. K.: Conformational changes and regulation of glutamate dehydrogenase activity. Biochem. J. **98**, 105—111 (1966).
Bernardi, G., Timasheff, S. N.: Optical rotatory dispersion and circular dichroism properties of yeast mitochondrial DNA's. J. molec. Biol. **48**, 43—52 (1970).
Bertland, L. H., Kaplan, N. O.: Studies on the conformation of the multiple forms of chicken heart aspartate aminotransferase. Biochemistry **9**, 2653—2665 (1970).
Bertland, A. U., Kalckar, H. M.: Reversible changes of ordered polypeptide structures in oxidized and reduced epimerase. Proc. Nat. Acad. Sci. US **61**, 629—635 (1968).
Bewley, T. A., Li, C. H.: Human pituitary growth hormone. XV. The optical rotatory dispersion in acidic, neutral, and basic solutions. Biochim. et Biophys. Acta **140**, 201—207 (1967).
Bewley, T. A., Li, C. H.: Circular dichroism studies on human chorionic somatotropin. Arch. Biochem. Biophys. **144**, 589—595 (1971).
Bewley, T. A., Sairam, M. R., Li, C. H.: Circular dichroism of ovine interstitial cell stimulating hormone and its subunits. Biochemistry **11**, 932—936 (1972).
Beychok, S.: Effect of ligands on the optical rotatory dispersion of hemoglobin. I. Ferrihemoglobin cyanide, ferrihemoglobin azide, ferrihemoglobin hydroxide, and carbonmonoxyhemoglobin. Biopolymers **2**, 575—584 (1964).
Beychok, S.: Circular dichroism of biological macromolecules. Science **154**, 1288—1299 (1966).
Beychok, S.: Circular dichroism of poly-α-amino acids and proteins. In: Poly-α-amino acids, protein models for conformational studies (ed. G. D. Fasman), pp. 293—337. New York: M. Dekker, Inc. 1967.
Beychok, S., Kabat, A.: Optical activity and conformation of carbohydrates. I. Optical rotatory dispersion studies on immunochemically reactive amino sugars and their glycosides, milk oligosaccharides of glucose, and blood group substances. Biochemistry **4**, 2565—2574 (1965).
Beychok, S., Armstrong, J. McD., Lindblow, C., Edsall, J. T.: Optical rotatory dispersion and circular dichroism of human carbonic anhydrases B and C. J. Biol. Chem. **241**, 5150 to 5160 (1966).
Beychok, S.: Rotatory dispersion and circular dichroism. Ann. Rev. Biochem. **37**, 437—462 (1968).
Bezkorovainy, A., Springer, G. F., Hotta, K.: Some physical properties of human NN and Me-Vg blood-group antigens. Biochim. et Biophys. Acta **115**, 501—504 (1966).
Bezkorovainy, A., Springer, G. F., Desai, P. R.: Physicochemical properties of the eel anti-human blood-group H(0) antibody. Biochemistry **10**, 3761—3764 (1971).
Biltonen, R., Lumry, R., Madison, V., Parker, H.: Studies of the chymotrypsinogen family. III. Optical rotatory dispersion of α-chymotrypsin. Proc. Nat. Acad. Sci. US **54**, 1018 to 1025 (1965).
Biot, J. B. (1817), quoted from Lowry, T. M.: Optical rotatory power, p. 20, 1935. New York: Dover Publ. (reprinted 1964).
Björk, I., Karlsson, F. A., Berggård, I.: Independent folding of the variable and constant halves of a lambda immunoglobulin light chain. Proc. Nat. Acad. Sci. US **68**, 1707—1710 (1971).
Blake, C. C. F., Koenig, D. F., Mair, G. A., North, A. C. T., Phillips, D. C., Sarma, V. R.: Structure of hen egg-white lysozyme. A three-dimensional Fourier synthesis at 2 Å resolution. Nature **206**, 757—763 (1965).

BLOUT, E. R.: Polypeptides and proteins. In: Optical rotatory dispersion. Applications to organic chemistry (ed. C. D. DJERASSI), pp. 238—273. New York-Toronto-London: McGraw-Hill Co. 1960.

BLOUT, E. R.: Extrinsic and intrinsic Cotton effects in polypeptides and proteins. Biopolymers, Symp. No 1, 397—408 (1964).

BLOUT, E. R., SHECHTER, E.: A new technique for producing oriented synthetic polypeptides. Some initial results. Biopolymers 1, 565—568 (1963).

BLOUT, E. R., SCHMIER, I., SIMMONS, N. S.: New Cotton effects in polypeptides and proteins. J. Am. Chem. Soc. 84, 3193—3194 (1962).

BLOUT, E. R., CARVER, J. P., GROSS, J.: Intrinsic Cotton effects in collagen and poly-L-proline. J. Am. Chem. Soc. 85, 644—645 (1963).

BLOW, D. M., STEITZ, T. A.: X-ray diffraction studies on enzymes. Ann. Rev. Biochem. 39, 63—100 (1970).

BLUNDELL, T. L., CUTFIELD, J. F., CUTFIELD, S. M., DODSON, E. J., DODSON, G. G., HODGKIN, D. C., MERCOLA, D. A., VIJAYAN, M.: Atomic positions in rhombohedral 2 zinc insulin crystals. Nature 231, 506—511 (1971).

BOLOTINA, I. A., VOLKENSTHEIN, M. V., ZAVODSKII, P., MARKOVICH, D. S.: Polarimetricheskoe izsledovanie konformacionnykh izmeneniy D-gliceraldegid-3-fosfat degidrogenazy (Polarimetric studies on conformational changes of D-glyceraldehyde-3-phosphate dehydrogenase). Biokhimiya 31, 649—653 (1966).

BOLOTINA, I. A., MARKOVICH, D. S., VOLIKENSTHEIN, M. V., ZAVODSKII, P.: Izuchenie konformacii laktatdegidrogenazy i yee kataliticheskoi aktivnosti (Study of the conformation of lactic dehydrogenase and its catalytical activity). Molek. Biologiya 1, 231—239 (1967).

BONNER, J., Ts'O, P.: The nucleohistones. San Francisco: Holden-Day Publ. 1964.

BOSSA, F., BRUNORI, M., BATES, G. W., ANTONINI, E., FASELLA, P.: Studies on hemerythrin, II. Circular dichroism and optical rotatory dispersion of hemerythrin from *Sipunculus nudus*. Biochim. et Biophys. Acta 207, 41—48 (1970).

BOUCHER, L. J., CRESPI, H. L.: Optical rotatory dispersion of phycocyanin. Biochemistry 5, 3796—3802 (1966).

BRADBURY, E. M., DOWIE, A. R., ELLIOTT, A., HANBY, W. E.: The stability of the screw sense of the α-helix in poly-β-benzyl-L-aspartate. Proc. Roy. Soc. London A. 259, 110—128 (1960).

BRADBURY, E. M., CRANE-ROBINSON, C., PHILLIPS, D. M. P., JOHNS, E. W., MURRAY, K.: Conformational investigation of histones. Nature 205, 1315—1316 (1965).

BRADY, A. H., BEYCHOK, S.: Optical activity and conformation studies on pig heart lipoamide dehydrogenase. J. Biol. Chem. 244, 4634—4637 (1969).

BRAHMS, J., BRAHMS, S.: Circular dichroism of nucleic acids, in: Fine structure of proteins and nucleic acids (eds. G. D. FASMAN and S. N. TIMASHEFF), pp. 191—270. New York: M. Dekker Inc. 1970.

BRAHMS, J.: Conformation et stabilité des oligonucléotides et des acides nucléiques. J. chim. phys. 65, 105—113 (1968).

BRAND, E., ERLANGER, B.: The optical rotation of peptides. J. Am. Chem. Soc. 72, 3314—3315 (1960).

BRESLOW, E., BEYCHOK, S., HARDMAN, K. D., GURD, F. R. N.: Relative conformations of sperm whale metmyoglobin and apomyoglobin in solution. J. Biol. Chem. 240, 304—309 (1965).

BRESLOW, E., AANNING, H. L., ABRASH, L., SCHMIR, M.: Physical and chemical properties of the bovine neurophysins. J. Biol. Chem. 246, 5179—5188 (1971).

BROWN, F. R. III, CARVER, J. P., BLOUT, E. R.: Low temperature circular dichroism of poly-(glycyl-L-prolyl-L-alanine). J. molec. Biol. 39, 307—313 (1969).

BROWN, W. V., LEVY, R. I., FREDRICKSON, D. S.: Further characterization of apolipoproteins from the human plasma very low density lipoproteins. J. Biol. Chem. 245, 6588—6594 (1970).

BUDZYNSKI, A. Z.: Difference in conformation of fibrinogen degradation products as revealed by hydrogen exchange and spectropolarimetry. Biochim. et Biophys. Acta 229, 663—671 (1971).

BUSCH, H.: Histones and other nuclear proteins. New York: Academic Press 1965.

BUSH, C. A., TINOCO, I.: Calculation of the optical rotatory dispersion of dinucleoside phosphates. J. Mol. Biol. 23, 601—614 (1967).

BUERER, T., GUENTHARD, H. H.: Spektropolarimetrie II. Empfindlichkeit von photoelektrischen Polarimetern. Helv. Chim. Acta 43, 810—825 (1960).

CALDWELL, D. J., EYRING, H.: Optical rotation. Ann. Rev. Phys. Chem. 15, 281—310 (1964).

CALDWELL, D. J., EYRING, H.: The theory of optical activity. New York-London-Sydney-Toronto: Wiley-Interscience 1971.

CALLAGHAN, P., MARTIN, N. H.: The relation of the rotatory dispersion behavior of human serum albumin to its conformation. Biochem. J. 83, 144—151 (1962).

CAROLL, B., BLEI, I.: Measurement of optical activity: New approaches. Science 142, 200—208 (1963).

CARVER, J. P., SHECHTER, E., BLOUT, E. R.: Analysis of the optical rotatory dispersion of polypeptides and proteins. IV. A digital computer analysis for the region 190—600 nm. J. Am. Chem. Soc. 88, 2550—2561 (1966).

CARY, H., HAWES, R. C., HOOPER, P. B., DUFFIELD, J. J., GEORGE, K. P.: A recording spectropolarimeter. Appl. Optics 3, 329—337 (1964).

CASSIM, J. Y., YANG, J. T.: A computerized calibration of the circular dichrometer. Biochemistry 8, 1947—1951 (1969).

CASSIM, J. Y., YANG, J. T.: Critical comparison of experimental optical activity of helical polypeptides and the predictions of the molecular exciton model. Biopolymers 9, 1475—1502 (1970).

CATHOU, R. E., HAMMES, G. G., SCHIMMEL, P. R.: Optical rotatory dispersion of ribonuclease and ribonuclease-nucleotide complexes. Biochemistry 4, 2687—2690 (1965).

CATHOU, R. E., KULCZYCKI, A., JR., HABER, E.: Structural features of γ-immunoglobulin, antibody, and their fragments. Biochemistry 7, 3958—3964 (1968).

CHAO, L. P., EINSTEIN, E. R.: Physical properties of the bovine encephalitogenic protein; molecular weight and conformation. J. Neurochem. 17, 1121—1132 (1970).

CHENG, P.-Y.: β-Forming protein chain in a globular enzyme, deoxyribonuclease I. Proc. Nat. Acad. Sci. US 55, 1535—1538 (1966).

CHENG, P.-Y.: Optical rotatory dispersion, tryptophan location, and base distribution in tobacco mosaic virus. Biochemistry 7, 3367—3373 (1968).

CHOU, P. Y., WELLS, M., FASMAN, G. D.: Conformational studies on copolymers of hydroxypropyl-L-glutamine and L-leucine. Circular dichroism studies. Biochemistry 11, 3028—3043 (1972).

CHOWDHURY, F. H., JOHNSON, P.: Physico-chemical studies on bovine γ-globulin. Biochim. et Biophys. Acta 66, 218—228 (1963).

CHUNG, A. E., FRANSEN, J. S.: Oxidized triphosphopyridine nucleotide specific isocitrate dehydrogenase from *Azotobacter vinelandii*. Isolation and characterization. Biochemistry 8, 3175—3184 (1969).

CLOUGH, G. W.: The relationship between the optical rotatory powers and the relative configurations of optically active compounds. The influence of certain inorganic haloids on the optical rotatory powers of α-hydroxy acids, α-amino acids, and their derivatives. J. Chem. Soc. 113, 526—554 (1918).

CLUSKEY, J. E., WU, Y. V.: Optical rotatory dispersion, circular dichroism and infrared studies of wheat gluten proteins in various solvents. Cereal Chem. 48, 203—211 (1971).

COHEN, C., SZENT-GYORGYI, A. G.: Optical rotation and helical polypeptide chain configuration in α-proteins. J. Am. Chem. Soc. 79, 248 (1957).

COLEMAN, D. L., BLOUT, E. R.: The optical activity of the disulfide bond in L-cytine and some derivatives of L-cystine. J. Am. Chem. Soc. 90, 2404—2416 (1968).

COLEMAN, J. E.: Metallocarbonic anhydrases: optical rotatory dispersion and circular dichroism. Proc. Nat. Acad. Sci. US 59, 123—130 (1968).

COOMBES, J. D., KATCHALSKI, E., DOTY, P.: Optical rotation and configuration of poly-L-tyrosine. Nature 185, 534—535 (1960).

COTTON, A.: Recherches sur l'absorption et la dispersion de la lumière par les milieux doués du pouvoir rotatoire. Ann. chim. et phys. 8, 347—433 (1896).

CRABBÉ, P.: Optical rotatory dispersion and circular dichroism in organic chemistry. San Francisco: Holden-Day Publ. 1965.

CRAIG, J. C., ROY, S. K.: Optical rotatory dispersion and absolute configuration. I. α-Amino acids. Tetrahedron 21, 391—394 (1965).

CREWTHER, W. G., HARRAP, B. S.: The preparation and properties of a helix-rich fraction obtained by partial proteolysis of low sulfur S-carboxymethylkerateine from wool. J. Biol. Chem. 242, 4310—4319 (1967).

CUNNINGHAM, B. A., GOTTLIEB, P. D., KONIGSBERG, W. H., EDELMAN, G. M.: The covalent structure of a human γG-immunoglobulin. V. Partial amino acid sequence of the light chain. Biochemistry 7, 1983—1995 (1968).

DARNALL, D. W., BARELA, T. D.: β-Structure in glyceraldehyde-3-phosphate dehydrogenase. Biochim. et Biophys. Acta 236, 593—598 (1971).

DAVIES, D. R.: X-ray diffraction studies of macromolecules. Ann. Rev. Biochem. 36, 321—364 (1967).

DAVIDSON, B., FASMAN, G. D.: The conformational transitions of uncharged poly-L-lysine. α-helix-random coil-β-structure. Biochemistry 6, 1616—1629 (1967).

DICKERSON, R. E.: X-ray analysis and protein structure. In: The proteins; composition, structure, and function (ed. H. NEURATH), 2nd. ed., vol. II., pp. 603—778. New York and London: Academic Press 1964.

DICKERSON, R. E., GEIS, I.: The structure and action of proteins. New York-Evanston-London: Harper & Row 1969.

DICKERSON, R. E., TAKANO, T., EISENBERG, D., KALLAI, O. B., SAMSON, L., COOPER, A., MARGOLIASH, E.: Ferricytochrome c. General features of the horse and bonito proteins at 2.8 A resolution. J. Biol. Chem. 246, 1511—1533 (1971).

DI SABATO, G., KAPLAN, N. O.: The denaturation of lactic dehydrogenase. II. The effects of urea and salts. J. Biol. Chem. 240, 1072—1076 (1965).

DJERASSI, C.: Optical rotatory dispersion. Applications to organic chemistry. New York-Toronto-London: McGraw-Hill Publ. 1960.

DJERASSI, C., BUNNENBERG, E., ELDER, D. L.: Organic chemical applications of magnetic circular dichroism. In: The chemistry of natural products, vol. 7. pp. 57—90. (Plenary lectures presented at the 7th Internat. Symp. on Chemistry of Natural Products, Riga, 1970). London: Butterworths 1971.

DOI, E., JIRGENSONS, B.: Circular dichroism studies on the acid denaturation of γ-immunoglobulin G and its fragments. Biochemistry 9, 1066—1073 (1970).

DORRINGTON, K. J., ZARLENGO, M. H., TANFORD, C.: Conformational change and complementarity in the combination of H and L chains of immunoglobulin-G. Proc. Nat. Acad. Sci. US 58, 996—1003 (1967).

DORRINGTON, K. J., TANFORD, CH.: The optical rotatory dispersion of human γ-M-immunoglobulins and their subunits. J. Biol. Chem. 243, 4745—4749 (1968).

DORRINGTON, K. J., SMITH, B. R.: Conformational changes accompanying the dissociation and association of immunoglobulin subunits. Biochim. et Biophys. Acta 263, 70—81 (1972).

DORRINGTON, K. J., TANFORD, C.: Molecular size and conformation of immunoglobulins. Advances Immunology 12, 333—381 (1970).

DOSCHER, M. S., RICHARDS, F. M.: The activity of an enzyme in the crystalline state: ribonuclease S. J. Biol. Chem. 238, 2399—2406 (1963).

DOTY, P., GEIDUSCHEK, E. P.: Optical properties of proteins. In: The proteins (eds. H. NEURATH and K. BAILEY), vol. I, p. A, pp. 393—460. New York: Academic Press 1953.

DOTY, P., IMAHORI, K., KLEMPERER, E.: The solution properties and configuration of a polyampholytic polypeptide. Copoly-L-lysine-L-glutamic acid. Proc. Nat. Acad. Sci. US 44, 424—431 (1958).

DRATZ, E. A., CALVIN, M.: Substrate- and inhibitor-induced changes in the optical rotatory dispersion of aspartate transcarbamylase. Nature 211, 497—501 (1966).

DRENTH, J., JANSONIUS, J. N., KOEKOEK, R., SWEN, H. M., WOLTHERS, B. G.: Structure of papain. Nature 218, 929—932 (1968).

DRENTH, J., JANSONIUS, J. N., KOEKOEK, R., WOLTHERS, B. G.: Papain, X-ray structure. In: The enzymes, 3rd ed., vol. III (ed. P. D. BOYER), pp. 485—499. New York-London:Academic Press 1971.

DRUDE, P.: Lehrbuch der Optik, 2nd ed. 1900. Leipzig: S. Hirzel Verlag 1906.

DUCKWORTH, H. W., COLEMAN, J. E.: Physicochemical and kinetic properties of mushroom tyrosinase. J. Biol. Chem. 245, 1613—1625 (1970).

EDELHOCH, H., LIPPOLDT, R. E.: The properties of thyroglobulin. II. The effects of sodium dodecyl sulfate. J. Biol. Chem. 235, 1335—1340 (1960).

EDELHOCH, H., METZGER, H.: The properties of thyroglobulin. V. The properties of denatured thyroglobulin. J. Am. Chem. Soc. 83, 1428—1435 (1961).

EDELHOCH, H., LIPPOLDT, R. E.: Structural studies on polypeptide hormones. J. Biol. Chem. 244, 3876—3883 (1969).

EDELMAN, G. M., GALLY, J. A.: The nature of Bence Jones proteins. Chemical similarities to polypeptide chains of myeloma globulins and normal γ-globulins. J. Exp. Med. 116, 207—228 (1962).

EDELMAN, G. M.: The covalent structure of human γG-immunoglobulin. XI. Functional implications. Biochemistry 9, 3197—3205 (1970).

EDMONDSON, D. E., TOLLIN, G.: Chemical and physical characterization of the Shethna flavoprotein and apoprotein and kinetics and thermodynamics of flavin analog to the apoprotein. Biochemistry 10, 124—132 (1971).

EDSALL, J. T., MEHTA, S., MYERS, D. V., ARMSTRONG J. McD.: Structure and denaturation of human carbonic anhydrases in urea and guanidine hydrochloride solutions. Biochem. Z. 345, 9—36 (1966).

EMERSON, T. R., SWAN, R. J., ULBRICHT, T. L. V.: Optical rotatory dispersion of nucleic acid derivatives. VIII. The conformation of pyrimidine nucleosides in solution. Biochemistry 6, 843—850 (1967).

EPAND, R. M., SCHERAGA, H. A.: The influence of longe-range interactions on the structure of myoglobin. Biochemistry 7, 2864—2872 (1968).

EYRING, E. J., YANG, J. T.: Conformation of protein-polysaccharide complex from bovine nasal septum. J. Biol. Chem. 243, 1306—1311 (1968).

EYRING, H., LIU, H.-C., CALDWELL, D.: Optical rotatory dispersion and circular dichroism. Chem. Revs. 68, 525—540 (1968).

FASELLA, P., HAMMES, G. G.: Ultraviolet rotatory dispersion of aspartic aminotransferase. Biochemistry 4, 801—805 (1965).

FASMAN, G. D.: The relative stability of the α-helix in synthetic polypeptides. In: Polyamino acids, polypeptides, and proteins (ed. M. A. STAHMANN), pp. 221—224. Madison: University of Wisconsin Press 1962.

FASMAN, G. D.: Optical rotatory dispersion. In: Methods in enzymology (eds. S. P. COLOWICK and N. O. KAPLAN), vol. VI, pp. 928—957. New York and London: Academic Press 1963.

FASMAN, G. D.: Factors responsible for conformational stability. In: Poly-α-amino acids, protein models for conformational studies (ed. G. D. FASMAN), pp. 499—604. New York: M. Dekker, Inc. 1967.

FASMAN, G. D., POTTER, J.: The optical rotatory dispersion of two beta structures. Biochem. Biophys. Res. Communs. 27, 209—216 (1967).

FASMAN, G. D., FOSTER, R. J., BEYCHOK, S.: The conformational transition associated with the activation of chymotrypsinogen to chymotrypsin. J. Mol. Biol. 19, 240—253 (1966).

FASMAN, G. D., LINDBLOW, C., BODENHEIMER, E.: Conformational studies on synthetic poly-α-amino acids; factors influencing the stability of the helical conformation of poly-L-glutamic acid and copolymers of L-glutamic acid and L-leucine. Biochemistry 3, 155—166 (1964).

FASMAN, G. D., LINDBLOW, C., GROSSMAN, L.: The helical conformations of polycytidylic acid; studies on the forces involved. Biochemistry 3, 1015—1021 (1964).

FASMAN, G. D., LINDBLOW, C., SEAMAN, E.: Optical rotatory dispersion studies on the conformational stabilization forces of yeast soluble ribonucleic acid. J. Mol. Biol. 12, 630—640 (1965).

FASMAN, G. D., HOVING, H., TIMASHEFF, S.: Circular dichroism of polypeptide and protein conformations. Film studies. Biochemistry 9, 3316—3324 (1970).

FASMAN, G. D., SCHAFFHAUSEN, B., GOLDSMITH, L., ADLER, A.: Conformational changes associated with fl histone-DNA complexes. Circular dichroism studies. Biochemistry 9, 2814—2822 (1970).

FERMANDJIAN, S., FROMAGEOT, P., TISTCHENKO, A. M., LEICKNAM, J. P., LUTZ, M.: Angiotensin II conformations. Infrared and Raman studies. Eur. J. Biochem. 28, 174—182 (1972).

FINKELSTEIN, A. V., PTITSYN, O. B.: Statistical analysis of the correlation among amino acid residues in helical, β-structural and non-regular regions in globular proteins. J. molec. Biol. 62, 613—624 (1971).

FLORY, P. J.: Configurational statistics of polypeptide chains. In: Conformation of biopolymers (ed. G. N. RAMACHANDRAN), vol. 1, pp. 339—363. London-New York: Academic Press 1967.

FOSTER, J. F.: Plasma albumin. In: The plasma proteins (ed. F. W. PUTNAM), vol. I., pp. 179—239. New York and London: Academic Press 1960.

FRANK, E. H., VEROS, A. J.: Physical studies on proinsulin. Association behavior and conformation in solution. Biochem. Biophys. Res. Communs. 32, 155—160 (1968).

FRANK, B. H., PEKAR, A. H., VEROS, A. J., Ho, P. P. K.: Crystalline L-asparaginase from Escherichia coli B. J. Biol. Chem. 245, 3716—3724 (1970).

FRASER, R. D. B., SUZUKI, E.: Polypeptide chain conformation in feather keratin. J. Mol. Biol. 14, 279—282 (1965).

FRESNEL, A. (1824), quoted from LOWRY, T. M.: Optical rotatory power, p. 7, 1935. New York: Dover Publ. (reprinted) 1964.

FRETTO, L., STRICKLAND, E. H.: Effect of temperature upon the conformations of carboxypeptidase A (Anson), A_{γ}^{Leu}, A_{γ}^{Val}, and $A_{\alpha+\beta}$. Biochim. et Biophys. Acta 235, 473—488 (1971).

FRETTO, L., STRICKLAND, E. H.: Use of circular dichroism to study interactions of carboxypeptidase A (Anson) and $A_{\alpha+\beta}$ with substrates and inhibitors. Biochim. et Biophys. Acta 235, 489—502 (1971).

FRUTON, J. S.: Chemical aspects of protein synthesis. In: The proteins; composition, structure, and function (ed. H. NEURATH), 2nd ed., vol. I., pp. 190—310. New York and London: Academic Press 1963.

FUKUSHIMA, D.: Internal structure of soybean protein molecule (11 S protein) in aqueous solution. J. Biochem. (Tokyo) 57, 822—823 (1965).

FUNATSU, M., HARADA, Y., HAYASHI, K., JIRGENSONS, B.: Studies on activation of pepsinogen. I. Conformational change of pepsinogen in the process of activation in acid solution. Agr. Biol. Chem. (Tokyo) 35, 566—572 (1971).

GARBETT, K., GILLARD, R. D., KNOWLES, P. F., STANGROOM, J. E.: Cotton effects in plant ferredoxin and xanthine oxidase. Nature 215, 824—828 (1967).

GARNIER, J.: Conformation de la caséine β en solution. Analyse d'une transition thermique entre 5 et 40° C. J. Mol. Biol. 19, 586—590 (1966).

GHOSE, A. C., JIRGENSONS, B.: Circular dichroism studies on the variable and constant halves of $\varkappa$-type Bence Jones proteins. Biochim. et Biophys. Acta, 251, 14—20 (1971).

GIDEZ, L. I.: Lipoproteins. In: The encyclopedia of biochemistry (eds. R. J. WILLIAMS and E. M. LANSFORD), pp. 493—496. New York-Amsterdam-London: Reinhold Publ. 1967.

GILL, T. J., III, DOTY, P.: Studies on synthetic polypeptide antigens. II. The immunochemical properties of a group of linear synthetic polypeptides. J. Biol. Chem. 236, 2677—2683 (1961).

GILLESPIE, J. M.: The isolation and properties of some soluble proteins from wool. Australian J. Biol. Sci. 16, 259—280 (1963).

GLAZER, A. N., SIMMONS, N. S.: A solvent-sensitive aromatic Cotton effect in lysozyme. J. Am. Chem. Soc. 87, 2287—2288 (1965).

GLAZER, M., SINGER, S. J.: Extrinsic Cotton effects characteristic of specific hapten-antibody interactions. Proc. Nat. Acad. Sci. US 68, 2477—2479 (1971).

GLAZER, M., SINGER, S. J.: Circular dichroism and the conformations of membrane proteins. Biochemistry 10, 1780—1787 (1971 b).

GOLDSACK, D. E.: Relation of amino acid composition and the Moffitt parameters to the secondary structure of proteins. Biopolymers 7, 299—313 (1969).

GOTTSCHALK, A., GRAHAM, E. R. B.: The basic structure of glycoproteins. In: The proteins (ed. H. NEURATH), 2nd ed., Vol. IV, pp. 95—151. New York-London: Academic Press 1966.

GORBUNOFF, M. J.: Exposure of tyrosine residues in proteins. III. The reaction of cyanuric fluoride of N-acetylimidazole with ovalbumin, chymotrypsinogen, and trypsinogen. Biochemistry 8, 2591—2598 (1969).

GORBUNOFF, M. J.: On the role of hydroxyl-containing amino acids in trypsin inhibitors. Biochim. et Biophys. Acta 221, 314—325 (1970).

GOTTO, A. M., LEVY, R. I., FREDRICKSON, D. S.: Observations on the conformation of human β-lipoprotein: evidence for the occurrence of β-structure. Proc. Nat. Acad. Sci. US 60, 1436—1441 (1968).

GOULD, H. J., GILL, T. J., III, DOTY, P.: The conformation and the hydrogen ion equilibrium of normal rabbit γ-globulin. J. Biol. Chem. 239, 2842—2851 (1964).

GRATZER, W. B., LOWEY, S.: Effect of substrate on the conformation of myosin. J. Biol. Chem. 244, 22—25 (1969).

GRATZER, W. B.: Ultraviolet absorption spectra of polypeptides. In: Poly-α-amino acids protein models for conformational studies (ed. G. D. FASMAN), pp. 177—238. New York: M. Dekker, Inc. 1967.

GRATZER, W. B., BEAVEN, G. H., RATTLE, H. W. E., BRADBURY, E. M.: A conformational study of glucagon. Eur. J. Biochem. 3, 276—283 (1968).

GRATZER, W. B., HILL, L. R., OWEN, R. J.: Circular dichroism of DNA. Eur. J. Biochem. 15, 209—214 (1970).

GREEN, N. M., MELAMED, M. D.: Optical rotatory dispersion, circular dichroism, and far ultraviolet spectra of avidin and streptavidin. Biochem. J. 100, 614—621 (1966).

GREENFIELD, N., DAVIDSON, B., FASMAN, G. D.: The use of computed optical rotatory dispersion curves for the evaluation of protein conformation. Biochemistry 6, 1630—1635 (1967).

GREENFIELD, N., FASMAN, G. D.: Computed circular dichroism spectra for the evaluation of protein conformation. Biochemistry 8, 4108—4116 (1969).

GREENSTEIN, J. P., WINITZ, M.: Chemistry of the amino acids, vol. I, pp. 3—244. New York: J. Wiley & Sons 1961.

GRIZZUTI, K., PERLMANN, G. E.: Conformation of the phosphoprotein phosvitin. J. Biol. Chem. 245, 2573—2578 (1970).

GROS, F.: The cell machinery. In: Molecular biophysics (eds. B. PULLMAN and M. WEISSBLUTH), pp. 1—80. New York and London: Academic Press 1965.

GROSS, J.: Comparative biochemistry of collagen. In: Comparative biochemistry (eds. M. FLORKIN and H. S. MASON), vol. 5, pp. 307—346. New York-London: Academic Press 1963.

GUHA, A., ENGLARD, S., LISTOWSKY, I.: Beef heart malic dehydrogenase. VII. Reactivity of sulfhydryl groups and conformation of the supernatant enzyme. J. Biol. Chem. 243, 609—615 (1968).

GUTTE, B., MERRIFIELD, R. B.: The synthesis of ribonuclease A. J. Biol. Chem. 246, 1922—1941 (1971).

HAMAGUCHI, K., MIGITA, S.: Optical rotatory and ultraviolet spectral properties of Bence Jones proteins. J. Biochem. (Tokyo) 56, 512—521 (1964).

HAMAGUCHI, K., IKEDA, K., LEE, C.-Y.: Optical rotatory dispersion and circular dichroism of neurotoxins isolated from the venom of Bungarus multicinctus. J. Biochem. (Tokyo) 64, 503—506 (1968).

HAMAGUCHI, K., IKEDA, K., YOSHIDA, C., MORITA, Y.: Circular dichroism of japanese radish peroxidase a. J. Biochem. (Tokyo) 66, 191—201 (1969).

HARMSEN, B. J. M., VAN DAM, A. F., VAN OS, G. A. J.: The conformation of eye lens proteins studied by means of optical rotatory dispersion. Biochim. et Biophys. Acta 126, 540—550 (1966).

HARRINGTON, W. F., VON HIPPEL, P. H.: The structure of collagen and gelatin. Advances in Protein Chem. 16, 1—138 (1961).

HARRINGTON, W. F., SELA, M.: Studies on the structure of poly-L-proline in solution. Biochim. et Biophys. Acta 27, 24—41 (1958).

HARRINGTON, W. F., JOSEPHS, R., SEGAL, D. M.: Physical chemical studies on proteins and polypeptides. Ann. Rev. Biochem. 35, II, 599—650 (1966).

HARRISON, S. C., BLOUT, E. R.: Reversible conformational changes of myoglobin and apomyoglobin. J. Biol. Chem. 240, 299—303 (1965).

HARTLEY, B. S., SHOTTON, D. M.: Pancreatic elastase. In: The enzymes (ed. P. D. BOYER), 3rd ed., vol. 3, pp. 323—373. New York-London: Academic Press 1971.

HARTSUCK, J. A., LIPSCOMB, W. N.: Carboxypeptidase A. In: The enzymes (ed. P. D. BOYER), 3rd ed., vol. 3, pp. 1—56. New York-London: Academic Press 1971.

HASHIZUME, H., SHIRAKI, M., IMAHORI, K.: Study of circular dichroism of proteins and polypeptides in relation to their backbone and side chain conformations. J. Biochem. (Tokyo) 62, 543—551 (1967).

HAUROWITZ, F.: Ionenstruktur, Löslichkeit und Flockung der Proteine. Ein Beitrag zur Systematik der Proteine. Kolloid-Z. 74, 208—218 (1936).

HAVSTEEN, B. H.: Studies of conformational changes in glyceraldehyde-3-phosphate dehydrogenase accompanying its catalytic action. Acta Chem. Scand. 19, 1643—1651 (1965).

HEINS, J. N., SURIANO, J. R., TANIUCHI, H., ANFINSEN, C. B.: Characterization of a nuclease produced by Staphylococcus aureus. J. Biol. Chem. 242, 1016—1020 (1967).

HELLER, W.: Polarimetry. In: Technique of Organic chemistry (ed. A. WEISSBERGER), vol. I. Physical methods of organic chemistry, part III, 3rd ed., pp. 2147—2333. New York-London: Interscience Publ. 1960.

HERSKOVITS, T. T.: On the conformation of caseins. Optical rotatory properties. Biochemistry 5, 1018—1026 (1966).

HERSKOVITS, MESCANTI, L.: Conformation of proteins and polypeptides. II. Optical rotatory dispersion and conformation of the milk proteins and other proteins in organic solvents. J. Biol. Chem. 240, 639—644 (1965).

HEWITT, L. F.: Optical rotatory power and dispersion of proteins. Biochem. J. 21, 216—224 (1927).

HIBINO, Y., SAMEJIMA, T., KAJIYAMA, S., NOSOH, Y.: On the conformation of porcine ceruloplasmin. Arch. Biochem. Biophys. 130, 617—623 (1969).

HILTNER, W. A., HOPFINGER, A. J., WALTON, A. G.: Helix-coil controversy for polyamino acids. J. Am. Chem. Soc. 94, 4324—4327 (1972).

HNILICA, L. S.: Proteins in the cell nucleus. Progr. Nucleic Acid Res. and Molecular Biology 7, 25—106 (1967).

HNILICA, L. S.: The structure and biological functions of histones. Cleveland, Ohio: Chemical Rubber Publ. 1972.

HOLLEY, R. W., APGAR, J., EVERETT, G. A., MADISON, J. T., MARQUISEE, M., MERRILL, S. H., PENSWICK, J. R., ZAMIR, A.: Structure of a ribonucleic acid. Science 147, 1462—1465 (1965).

HOLZWARTH, G., DOTY, P.: The ultraviolet circular dichroism of polypeptides. J. Am. Chem. Soc. 87, 218—228 (1965).

HOOKER, T. M., SCHELLMAN, J. A.: Optical activity of aromatic chromophores. I. o-, m-, and p-tyrosine. Biopolymers 9, 1319—1348 (1970).

HORWITZ, J., STRICKLAND, E. H., BILLUPS, C.: Analysis of the vibrational structure in the near ultraviolet circular dichroism and absorption spectra of tyrosine derivatives and ribonuclease A at 77° K. J. Am. Chem. Soc. 92, 2119—2129 (1970).

HRKAL, Z., VODRAZKA, Z.: A study of the conformation of human globin in solution by optical methods. Biochim. et Biophys. Acta 133, 527—534 (1967).

HUBER, R., KUKLA, D., RÜHLMANN, A., EPP, O., FORMANEK, H.: The basic trypsin inhibitor of bovine pancreas. Naturwissenschaften 57, 389—392 (1970).

ICHISHIMA, E., YOSHIDA, F.: Conformation of aspergillopeptidase A in aqueous solution. Biochim. et Biophys. Acta 128, 130—135 (1966).

ICHISHIMA, E., YOSHIDA, F.: Conformation of aspergillopeptidase A in aqueous solution. II. Ultraviolet optical rotatory dispersion of aspergillopeptidase A. Biochim. et Biophys. Acta 147, 341—346 (1967).

IIZUKA, E., YANG, J. T.: Optical rotatory dispersion of L-amino acids in acid solution. Biochemistry 3, 1519—1524 (1964).

IIZUKA, E., YANG, J. T.: Optical rotatory dispersion and circular dichroism of the β-form of silk fibroin in solution. Proc. Nat. Acad. Sci. US 55, 1175—1182 (1966).

IIZUKA, E.: Species specificity of the conformation of silk fibroin in solution. Biochim. et Biophys. Acta 160, 454—463 (1968).

IIZUKA, E., YANG, J. T.: The disordered and β conformations of silk fibroin in solution. Biochemistry 7, 2218—2228 (1968).

IKEDA, K., HAMAGUCHI, K., YAMAMOTO, M., IKENAKA, T.: Circular dichroism and optical rotatory dispersion of trypsin inhibitors. J. Biochem. (Tokyo) 63, 521—531 (1968).

IKEDA, K., HAMAGUCHI, K., MIGITA, S.: Circular dichroism of Bence Jones proteins and immunoglobulins G. J. Biochem. (Tokyo) 63, 654—660 (1968).

IKEDA, K., HAMAGUCHI, K.: A tryptophyl circular dichroism band at 305 nm of hen egg-white lysozyme. J. Biochem. (Tokyo) 71, 265—273 (1972).

IMAHORI, K.: β-Conformation in γ-globulin. Biopolymers 1, 563—565 (1963).

IKEDA, K., HAMAGUCHI, K., MIWA, S., NISHINA, T.: Circular dichroism of human lysozyme. J. Biochem. (Tokyo) 71, 371—378 (1972).

IMANISHI, A., MOMOTANI, Y., ISEMURA, T.: The interaction of detergents with proteins. The effect of detergents on the conformation of *Bacillis subtilis* α-amylase and Bence Jones protein. J. Biochem. (Tokyo) 57, 417—429 (1965).

INOUE, S., ANDO, T.: Optical rotatory dispersion of nucleoclupeine and other model nucleoproteins. Biochem. Biophys. Res. Communs. 32, 501—506 (1968).

INSKEEP, W. H., MILES, D. W., EYRING, H.: Circular dichroism of nucleoside derivatives. VIII. Coupled oscillator calculations of molecules with fixed structure. J. Am. Chem. Soc. 92, 3866—3872 (1970).

JAENICKE, R.: Zur Bestimmung des Molekulargewichts und der molekularen Struktur differenter und multipler Formen von Laktatdehydrogenase. II. Isozyme vom Herzmuskel-Typ. Biochim. et Biophys. Acta 85, 186—201 (1964).

JI, T. H., URRY, D. W.: Correlation of light scattering and absorption flattening effects with distortions in the circular dichroism patterns of mitochondrial membrane fragments. Biochem. biophys. Res. Commun. 34, 404—411 (1969).

JIRGENSONS, B.: (1) Optical rotation and viscosity of native and denatured proteins. X. Further studies on optical rotatory dispersion. Arch. Biochem. Biophys. 74, 57—69 (1958).

JIRGENSONS, B.: (2) Optical rotation and viscosity of native and denatured proteins. XI. Relationships between rotatory dispersion, ionization and configuration. Arch. Biochem. Biophys. 74, 70—83 (1958).

JIRGENSONS, B. The optical rotatory dispersion of Bence Jones proteins. Arch. Biochem. Biophys. 85, 89—96 (1959).

JIRGENSONS, B.: Optical rotatory dispersion of some pituitary hormones. Arch. Biochem. Biophys. 91, 123—129 (1960).

JIRGENSONS, B.: (1) Optical rotatory dispersion of globular proteins. Tetrahedron 13, 166—175 (1961).

JIRGENSONS, B.: (2) Effect of detergents on the conformation of proteins. I. An abnormal increase of the optical rotatory dispersion constant. Arch. Biochim. Biophys. 94, 59—67 (1961).

JIRGENSONS, B.: (3) Glutamic acid dehydrogenase — a protein of unusual conformation. J. Am. Chem. Soc. 83, 3161 (1961).

JIRGENSONS, B.: (2) Optical rotation and viscosity of native and denatured proteins. XIV. Effect of surface active anions on serum γ-globulin and trypsin inhibitor. Makromol. Chem. 51, 137—147 (1962).

JIRGENSONS, B.: (3) Natural organic macromolecules. Oxford-London-New York-Paris: Pergamon Press 1962.

JIRGENSONS, B.: Optical rotatory dispersion and conformation of various globular proteins. J. Biol. Chem. 238, 2716—2722 (1963).

JIRGENSONS, B.: Study of the optical rotatory dispersion of proteins with improved spectropolarimetric techniques. Makromol. Chem. 72, 119—130 (1964).

JIRGENSONS, B.: The Cotton effects in the optical rotatory dispersion of proteins as new criteria of conformation. J. Biol. Chem. 240, 1064—1071 (1965).

JIRGENSONS, B.: (1) Optical rotatory dispersion of nonhelical proteins. J. Biol. Chem. 241, 147—152 (1966).

JIRGENSONS, B.: (2) Effect of anionic detergents on the optical rotatory dispersion of proteins. J. Biol. Chem. 241, 4855—4860 (1966).

JIRGENSONS, B.: (3) Classification of proteins according to conformation. Makromol. Chem. 91, 74—86 (1966).

JIRGENSONS, B.: (1) The far-ultraviolet Cotton effects and conformation of ribonuclease and pepsin. J. Am. Chem. Soc. 89, 5979—5981 (1967).

JIRGENSONS, B.: (2) Effects of n-propyl alcohol and detergents on the optical rotatory dispersion of α-chymotrypsinogen, β-casein, histone fraction F1, and soybean trypsin inhibitor. J. Biol. Chem. 242, 912—918 (1967).

JIRGENSONS, B., HNILICA, L. S.: The conformational changes of calf thymus histone fractions as determined by the optical rotatory dispersion. Biochim. et Biophys. Acta 109, 241—249 (1965).

JIRGENSONS, B., HNILICA, L. S., CAPETILLO, S. C.: Viscosity and conformation of calf thymus histones. Makromol. Chem. 97, 216—224 (1966).

JIRGENSONS, B., SAINE, S., ROSS, D. L.: The ultraviolet rotatory dispersion and conformation of Bence Jones proteins. J. Biol. Chem. 241, 2314—2319 (1966).

JIRGENSONS, B., SPRINGER, G. F.: Conformation of blood-group and virus receptor glycoproteins from red cells and secretions. Science 162, 365—367 (1968).

JIRGENSONS, B.: Circular dichroism of proteins of known and unknown conformations. Biochim. et Biophys. Acta 200, 9—17 (1970).

JIRGENSONS, B.: Reconstructive denaturation of proteins by sodium dodecyl sulfate. Circular dichroism of soybean trypsin inhibitor, papain, and a mold protease. Makromol. Chem. 158, 1—8 (1972).

JIRGENSONS, B., WARD, D. N.: Circular dichroism of ovine luteinizing hormone and its subunits. Texas Repts. Biol. and Med. 28, 553—559 (1970).

JIRGENSONS, B., KAWABATA, M., CAPETILLO, S.: Circular dichroism of soybean trypsin inhibitor and its derivatives. Makromol. Chem. 125, 126—135 (1969).

JIRGENSONS, B., CAPETILLO, S.: Effect of sodium dodecyl sulfate on circular dichroism of some nonhelical proteins. Biochim. et Biophys. Acta 214, 1—5 (1970).

JIRGENSONS, B., SPRINGER, G. F., DESAI, P. R.: Comparative circular dichroism studies of individual eel and human antibodies. Comp. Biochem. Physiol. 34, 721—725 (1970).

JOHNS, E. W., BUTLER, J. A. V.: Further fractionation of histones from calf thymus. Biochem. J. 82, 15—18 (1962).

JOHNSON, G. F., GRAVES, D. J.: Circular dichroism and optical rotatory dispersion of glycogen phosphorylase. Biochemistry 5, 2906—2909 (1966).

JOHNSON, G. F., PHILIP, G., GRAVES, D. J.: Dinitrophenylation of glycogen phosphorylase. II. Circular dichroism of the modified enzyme. Biochemistry 7, 2101—2105 (1968).

JOHNSON, W. C.: A circular dichroism spectrometer for the vacuum ultraviolet. Rev. Sci. Instr. 42, 1283—1286 (1971).

JOHNSON, W. C., TINOCO, I.: Circular dichroism of polypeptide solutions in the vacuum ultraviolet. J. Am. Chem. Soc. 94, 4389—4390 (1972).

JOLY, M.: A physico-chemical approach to the denaturation of proteins. London-New York: Academic Press 1965.

JONES, H. A., LERMAN, S.: Optical rotatory dispersion and circular dichroism studies on ocular lens proteins. Canad. J. Biochem. and Physiol. 49, 426—430 (1971).

KABAT, E. A.: Unique features of the variable regions of the Bence Jones proteins and their possible relation to antibody complementarity. Proc. Nat. Acad. Sci. US 59, 613—619 (1968).

KANAREK, L., HILL, R. L.: The preparation and characterization of fumarase from swine heart muscle. J. Biol. Chem. 239, 4202—4206 (1964).

KANNAN, K. K., LILJAS, A., WAARA, I., BERGSTEN, P. C., LÖVGREN, S., STRANDBERG, B., BENGTSSON, U., CARLBOM, U., FRIDBORG, K., JÄRUP, L., PETEF, M.: Crystal structure of human erythrocyte carbonic anhydrase c. VI. The three-dimensional structure at high resolution in relation to other mammalian carbonic anhydrases. Cold Spring Harbor Symposia Quant. Biol. 36, 221—231 (1972).

KARTHA, G., BELLO, J., HARKER, D.: Tertiary structure of ribonuclease. Nature 213, 862—865 (1967).

KASARDA, D. D., BERNARDIN, J. E., GAFFIELD, W.: Circular dichroism and optical rotatory dispersion of α-gliadin. Biochemistry 7, 3950—3957 (1968).

KATCHALSKI, E., SELA, M.: Synthesis and chemical properties of poly-α-amino acids. Advances in Protein Chem. 13, 243—492 (1958).

KATCHALSKI, E., SELA, M., SILMAN, H. I., BERGER, A.: Polyamino acids as protein models. In: The proteins; composition, structure, and function (ed. H. NEURATH), 2nd ed., vol. II, pp. 405—602. New York-London: Academic Press 1964.

KATO, H., SUZUKI, T., IKEDA, K., HAMAGUCHI, K.: Physicochemical properties of bovine kinogen-II. J. Biochem. (Tokyo) 62, 591—598 (1967).

KAUZMANN, W.: The physical chemistry of proteins. Ann. Rev. Phys. Chem. 8, 413—438 (1957).

KAUZMANN, W.: Some factors in the interpretation of protein denaturation. Advances in Protein Chem. 14, 1—63 (1959).

KAUZMANN, W., WALTER, J. E., EYRING, H.: Theories of optical rotatory power. Chem. Revs. 26, 339—407 (1940).

KAWAUCHI, S., IWANAGA, S., SAMEJIMA, Y., SUZUKI, T.: Isolation and characterization of two phospholipase A's from the venom of Agkistrodon Halys Blomhoffii. Biochim. et Biophys. Acta 236, 142—160 (1971).

KAY, C. M.: The presence of β-structure in concanavalin A. FEBS Lett. 9, 78—80 (1970).

KAYNE, F. J., SUELTER, C. H.: The temperature-dependent conformational transitions of pyruvate kinase. Biochemistry 7, 1678—1684 (1968).

KENDREW, J. C.: Myoglobin and the structure of proteins. Science 139, 1259—1266 (1963).

KENDREW, J. C., DICKERSON, R. E., STRANDBERG, B. E., HART, R. G., DAVIES, D. R., PHILLIPS, D. C., SHORE, V. C.: Structure of myoglobin. A three-dimensional Fourier synthesis at 2 Å resolution. Nature 185, 422—427 (1960).

KENKARE, U. W., COLOWICK, S. P.: Reversible inactivation and dissociation of yeast hexokinase. J. Biol. Chem. 240, 4570—4584 (1965).

KESSLER, D., LEVINE, L., FASMAN, G. D.: Some conformational and immunological properties of a bovine brain acidic protein (S-100). Biochemistry 7, 758—764 (1968).

KINCAID, H. L., JIRGENSONS, B.: Circular dichroism of ϰ- and λ-type Bence Jones proteins at various pH values and temperatures. Biochim. et Biophys. Acta 271, 23—33 (1972).

KIRKWOOD, J. G.: The theory of optical rotatory power. J. Chem. Phys. 5, 479—491 (1937).

KIRSCHENBAUM, D. M.: Molar extinction coefficients and $E_{1cm}^{1\%}$ values for proteins at selected wavelengths of ultraviolet and visible region. In: Handbook of biochemistry, Selected data for molecular biology (ed. H. A. SOBER), 2nd ed., pp. C 71—C 98. Cleveland, Ohio: Chemical Rubber Publ. 1970.

KIRSCHENBAUM, D. M.: Atlas of protein spectra in the ultraviolet and visible regions. New York-Washington-London: IFI/Plenum Press 1972.

KITO, Y., AZUMA, M., MAEDA, Y.: Circular dichroism of squid rhodopsin. Biochim. et Biophys. Acta 154, 352—359 (1968).

KLEE, W. A., MUDD, S. H.: The conformation of ribonucleosides in solution. The effect of structure on the orientation of the base. Biochemistry 6, 988—998 (1967).

KLEE, C. B.: Reversible polymerisation of histidine ammonia lyase. The role of SH groups in the activity and polymeric state of enzyme. J. Biol. Chem. 245, 3143—3152 (1970).

KLEIN, D., YAGUCHI, M., FOSTER, J. F., KOFFLER, H.: Conformational transitions in flagellins. I. Hydrogen ion dependency. J. Biol. Chem. 243, 4931—4935 (1968).

KLOTZ, I. M.: Protein subunits: a table. Science 155, 697—698 (1967).

KOTAKI, A., SUGIURA, N., YAGI, K.: Circular dichroic spectra of D-amino acid oxidase and its complexes. Biochim. et Biophys. Acta 151, 689—691 (1968).

KORNGUTH, S. E., PERRIN, J. H.: Circular dichroism and viscosimetric studies on a basic protein from pig brain. J. Neurochem. 18, 983—988 (1971).

KRAUT, J., HIGH, D. F., SIEKER, L. C.: Chymotrypsinogen; increased resolution and absolute configuration. Proc. Nat. Acad. Sci. US 51, 839—845 (1964).

KRAUT, J.: Chymotrypsinogen: X-ray structure. In: The enzymes (ed. P. D. BOYER), 3rd ed., vol. 3, pp. 165—183. New York-London: Academic Press 1971.

KRAUT, J.: Subtilisin: X-ray structure. In The enzymes (ed. P. D. BOYER), 3rd ed., vol. 3, pp. 547—560. New York-London: Academic Press 1971.

KRONMAN, M. J., BLUM, R., HOLMES, L. G.: Estimation of helicity of proteins from optical rotatory dispersion measurements. Biochem. Biophys. Res. Communs. 19, 227—232 (1965).

KRONMAN, M. J.: Similarity in backbone conformation of egg white lysozyme and bovine α-lactalbumin. Biochem. Biophys. Res. Communs. 33, 535—541 (1968).

LASKOWSKI, M., JR., SEALOCK, R. W.: Protein proteinase inhibitors — molecular aspects. In: The enzymes (ed. P. D. BOYER), 3rd ed., vol. 3, pp. 375—473. New York-London: Academic Press 1971.

LEACH, S. J., NEMETHY, SCHERAGA, H. A.: Intramolecular steric effects and hydrogen bonding in regular conformations of polyamino acids. Biopolymers 4, 887—904 (1966).

LEDERER, F.: β-structure in acetoacetate decarboxylase. Biochemistry 7, 2168—2172 (1968).

LEE, P. K. J., JIRGENSONS, B.: Conformational transitions of immunoglobulin fragment Fc(t) incited by alkyl sulfates of various hydrophobic chain lengths. Biochim. et Biophys. Acta 229, 631—641 (1971).

LI, L., SPECTOR, A.: The optical rotatory dispersion and circular dichroism of calf lens α-crystallin. J. Biol. Chem. 242, 3234—3236 (1967).

LI, C. H., STARMAN, B.: Molecular weight of sheep pituitary interstitial cell stimulating hormone. Nature 202, 291—292 (1964).

LI, L. K., SPECTOR, A.: The circular dichroism of β-poly-L-lysine. J. Am. Chem. Soc. 91, 220—222 (1969).

LIPSCOMB, W. N., HARTSUCK, J. A., QUIOCHO, F. A., REEKE, G. N.: The structure of carboxypeptidase A. IX. The X-ray diffraction results in the light of the chemical sequence. Proc. Nat. Acad. Sci. US 64, 28—35 (1969).

LISTOWSKY, I., FURFINE, C. S., BETHELL, J. J., ENGLARD, S.: Coenzyme induced changes in the optical rotatory dispersion properties of glyceraldehyde-3-phosphate dehydrogenase. J. Biol. Chem. 240, 4253—4258 (1965).

LISTOWSKY, I., AVIGAD, G., ENGLARD, S.: Conformational aspects of muramic acids. Analysis based on circular dichroism measurements. Biochemistry 9, 2186—2189 (1970).

LITMAN, G. W., GOOD, R. A., FROMMEL, D., ROSENBERG, A.: Conformational significance of the intrachain disulfide linkages in immunoglobulins. Proc. Nat. Acad. Sci. US 67, 1085—1092 (1970).

LIU, W. K., NAHM, H. S., SWEENEY, C. M., LAMKIN, W. M., BAKER, H. N., WARD, D. N.: The primary structure of ovine luteinizing hormone. I. The amino acid sequence of the reduced and S-aminoethylated S-subunit (LH-α). J. Biol. Chem. 247, 4351—4364 (1972).

LIU, W. K., NAHM, H. S., SWEENEY, C. M., HOLCOMB, G. N., WARD, D. N.: The primary structure of ovine luteinizing hormone. II. The amino acid sequence of the reduced, S-carboxymethylated A-subunit (LH-β). J. Biol. Chem. 247, 4365—4381 (1972).

LOW, B. W., LOVELL, F. M., RUDKO, A. D.: Prediction of the α-helical regions in proteins of known sequence. Proc. Nat. Acad. Sci. US 60, 1519—1526 (1968).

LLOYD, K. O., BEYCHOK, S., KABAT, E. A.: Immunochemical studies on blood groups. XXXIX. Optical rotatory dispersion and circular dichroism spectra of oligosaccharides from blood-group Lewis substance. Biochemistry 7, 3762—3765 (1968).

LOWRY, T. M.: Optical rotatory power. Original published by Longmans, Green and Co., London 1935. New York: Dover Publ. (reprinted) 1964.

LUTZ, O., JIRGENSONS, B.: Eine neue Methode der Zuteilung optisch-aktiver α-Aminosäuren zur Rechts- oder Linksreihe. I. Ber. deutsch. chem. Ges. **63**, 448—460 (1930).

LUTZ, O., JIRGENSONS, B.: Eine neue Methode der Zuteilung optisch-aktiver α-Aminosäuren zur Rechts- oder Linksreihe. II. Ber. deutsch. chem. Ges. **64**, 1221—1232 (1931).

LUDWIG, M. L., HARTSUCK, J. A., STEITZ, T. A., MUIRHEAD, H., COPPOLA, J. C., REEKE, G. N., LIPSCOMB, W. N.: The structure of carboxypeptidase A. IV. Preliminary results at 2.8 Å resolution, and a substrate complex at 6 Å resolution. Proc. Nat. Acad. Sci. US **57**, 511—514 (1967).

LUX, S. E., HIRZ, R., SHRAGER, R. I., GOTTO, A. M.: The influence of lipid on the conformation of human plasma high density apolipoproteins. J. Biol. Chem. **247**, 2598—2606 (1972).

MADISON, V., SCHELLMAN, J.: Optical activity of polypeptides and proteins. Biopolymers **11**, 1041—1076 (1972).

MAESTRE, M. F., TINOCO, I., JR.: Optical rotatory dispersion of viruses. J. Mol. Biol. **23**, 323—335 (1967).

MAGAR, M. E.: On the analysis of the optical rotatory dispersion of proteins. Biochemistry **7**, 617—620 (1968).

MALMSTROM, B. G.: Metal ion activation, metal binding, and sulfhydryl groups in native and denatured enolase from rabbit muscle. Arch. Biochem. Biophys. Suppl.1, 247—259 (1962).

MANNIK, M., KUNKEL, H. G.: Classification of myeloma proteins, Bence Jones proteins, and macroglobulins into two groups on the basis of common antigenic characters. J. Exp. Med. **116**, 859—878 (1962).

MARKUS, G.: Electrolytic reduction of the disulfide bonds of insulin. J. Biol. Chem. **239**, 4163—4170 (1964).

MARTIN, R. B.: Introduction to biophysical chemistry. New York-San Francisco-Toronto-London: McGraw-Hill Book Company 1964.

MASOTTI, L., URRY, D. W., KRIVACIC, J. R., LONG, M. M.: Circular dichroism of biological membranes. II. Plasma membranes and sarcotubular vesicles. Biochim. et Biophys. Acta **266**, 7—17 (1972).

MATTHEWS, B. W., SIGLER, P. B., HENDERSON, R., BLOW, D. M.: Three-dimensional structure of tosyl-α-chymotrypsin. Nature **214**, 652—656 (1967).

MATSUYAMA, A., TAGASHIRA, Y., NAGATA, C.: A circular dichroism study on the conformation of DNA in rat liver chromatin. Biochim. et Biophys. Acta **240**, 184—190 (1971).

McCUBBIN, W. D., KAY, C. M., OIKAWA, K.: Polarimetric studies on the interaction of muscle proteins with detergents. Biochim. et Biophys. Acta **126**, 597—600 (1966).

McMULLEN, D. W., JASKUNAS, S. R., TINOCO, I.: Application of matrix rank analysis to the optical rotatory dispersion of tobacco mosaic virus ribonucleic acid. Biopolymers **5**, 589—613 (1967).

McPHIE, P., GRATZER, W. B.: The optical rotatory dispersion of ribosomes and their constituents. Biochemistry **5**, 1310—1315 (1966).

MELOUN, B., FRIC, I., SORM, F.: Reduction of disulfide bridge Cys 14-Cys 38 of pancreatic trypsin inhibitor by β-mercaptoethanol: characterization of arising protein derivative by optical dispersion and circular dichroism measurements. Collection Czechoslov. Chem. Commun. **33**, 2299—2306 (1968).

MENENDEZ, C. J., HERSKOVITS, T. T.: Optical rotatory dispersion and circular dichroism studies on insulin and its trypsin-modified derivatives. Arch. Biochem. Biophys. **140**, 286—294 (1970).

MENKE, W.: Far ultraviolet circular dichroism and infrared absorption of thylakoids. Z. Naturforsch. **25 b**, 849—855 (1970).

MERCOLA, D. A., MORRIS, J. W. S., ARQUILLA, E. R., BROMER, W. W.: The ultraviolet circular dichroism of bovine insulin and desoctapeptide insulin. Biochim. et Biophys. Acta **133**, 224—232 (1967).

MIGITA, S., PUTNAM, F. W.: Antigenic relationships of Bence Jones proteins and normal human γ-globulin. J. Exp. Med. **117**, 81—104 (1963).

MIHALYI, E.: Physicochemical studies of bovine fibrinogen. III. Optical rotatory dispersion of native and denatured molecule. Biochim. et Biophys. Acta **102**, 487—499 (1965).

MILES, D. W., ROBINS, R. K., EYRING, H.: Optical rotatory dispersion, circular dichroism, and absorption studies on some naturally occurring ribonucleosides and related derivatives. Proc. Nat. Acad. Sci. US 57, 1138—1145 (1967).

MILES, D. W., INSKEEP, W. H., ROBINS, M. J., WINKLEY, M. W., ROBINS, R. K., EYRING, H.: Circular dichroism of nucleoside derivatives. IX. Vicinal effects on the circular dichroism of pyrimidine nucleosides. J. Am. Chem. Soc. 92, 3872—3881 (1970).

MILSTEIN, C., FRANGIONE, B., PINK, J. R. L.: Studies on the variability of immunoglobulin sequence. Cold Spring Harbor Symposia Quant. Biol. 32, 31—36 (1967).

MIRSKY, R., GEORGE, P.: Optical rotatory dispersion spectra of cytochrome c and polymers. Proc. Nat. Acad. Sci. US 56, 222—229 (1966).

MIRSKY, R.: Optical rotatory dispersion and spectral properties of yeast isocytochromes c. Biochemistry 6, 3671—3675 (1967).

MITCHELL, W. M., HASH, J. H.: The N,O-Diacetylmuramidase of Calaropsis species. II. Physical properties. J. Biol. Chem. 244, 17—21 (1969).

MIYAZAWA, T.: Molecular vibrations and structure of high polymers. II. Helical parameters of infinite polymer chains as functions of bond length, bond angles, and internal rotation angles. J. Polymer Sci. 55, 215—231 (1961).

MIZUSHIMA, S., SHIMANOUCHI, T.: Possible polypeptide configurations of proteins from the viewpoint of internal rotation potential. Advances in Enzymol. 23, 1—27 (1961).

MOFFITT, W.: Optical rotatory dispersion of helical polymers. J. Chem. Phys. 25, 467—478 (1956).

MOFFITT, W., MOSCOWITZ, A.: Optical activity in absorbing media. J. Chem. Phys. 30, 648—660 (1959).

MOFFITT, W.: YANG, J. T.: The optical rotatory dispersion of simple polypeptides. I. Proc. Nat. Acad. Sci. US 42, 596—603 (1956).

MOMMAERTS, W. F. H. M.: Ultraviolet circular dichroism in nucleic acid structural analysis. In: Methods in enzymology, vol. XII (B) (eds. L. GROSSMAN and K. MOLDAVE), pp. 302—329. New York: Academic Press 1968.

MORRIS, J. W. S., MERCOLA, D. A., ARQUILLA, E. R.: An analysis of the near ultraviolet circular dichroism of insulin. Biochim. et Biophys. Acta 160, 145—150 (1968).

MOSCOWITZ, A.: Theory and analysis of rotatory dispersion curves. In: Optical rotatory dispersion; applications to organic chemistry (ed. C. DJERASSI), pp. 150—177. New York-Toronto-London: McGraw-Hill Publ. 1960.

MURAKAMI, H.: Rotatory dispersion of protein and deoxyribose nucleic acid (DNA). J. Chem. Phys. 27, 1231—1237 (1957).

MURPHY, A. J.: Circular dichroism of the adenine and 6-mercaptopurine nucleotide complexes with actin. Biochemistry 10, 3723—3728 (1971).

MURRAY, A. C., OIKAWA, K., KAY, C. M.: Circular dichroism studies on native fetuin and some of its derivatives. Biochim. et Biophys. Acta 175, 331—338 (1969).

MYER, Y. P., HARBURY, H. A.: Optical rotatory dispersion of cytochrome c. Proc. Nat. Acad. Sci. US 54, 1391—1398 (1965).

MYER, Y. P.: Far ultraviolet circular dichroism spectra of cytochrom c. Biochim. et Biophys. Acta 154, 84—90 (1968).

MYER, Y. P., KING, T. E.: Effects of the oxidation state, ligands, detergents, and aging on the conformation of cytochrome oxidase. Biochem. Biophys. Res. Communs. 33, 43—48 (1968).

MYERS, D. V., EDSALL, J. T.: Optical rotatory dispersion of human carbonic anhydrases; Cotton effects and aromatic absorption bands. Proc. Nat. Acad. Sci. US 53, 169—177 (1965).

NAGY, B., LEHRER, S. S.: Circular dichroism of iron, copper, and zinc complexes of transferrin. Arch. Biochem. Biophys. 148, 27—36 (1972).

NAGY, B., STRZELECKA-GOLASZEWSKA, H.: Optical rotatory dispersion and circular dichroism spectra of G-actin. Arch. Biochem. Biophys. 150, 428—435 (1972).

NAKAGAWA, Y., PERLMANN, G. E.: Reduction and reoxidation of pepsinogen and pepsin. Arch. Biochem. Biophys. 140, 464—473 (1970).

NAKAGAWA, Y., CAPETILLO, S., JIRGENSONS, B.: Effect of chemical modification of lysine residues on the conformation of human immunoglobulin-G. J. Biol. Chem. 247, 5703—5708 (1972).

NEEDLEMAN, S. B. (ed.): Protein sequence determination. In: Molecular biology, biochemistry and biophysics (eds. W. KLEINZELLER, G. F. SPRINGER, H. G. WITTMANN), vol. 8. Berlin-Heidelberg-New York: Springer 1970.

NEMETHY, G., PHILLIPS, D. C., LEACH, S. J., SCHERAGA, H. A.: A second right-handed helix structure with the parameters of the Paulin-Corey α-helix. Nature 214, 363—365 (1967).

NOELKEN, M., REIBSTEIN, M.: Conformation of β-casein B. Arch. Biochem. Biophys. 123, 397—402 (1968).

NORTH, A. C. T., PHILLIPS, D. C.: X-ray studies of crystalline proteins. Progr. Biophys. Mol. Biol. 19, 5—132 (1969).

NOVAK, R. L., DOTY, P.: Optical rotatory dispersion and circular dichroism studies on Escherichia coli ribonucleic acid polymerase. Biochemistry 9, 1739—1743 (1970).

OIKAWA, K., KAY, C. M., McCUBBIN, W. D.: The ultraviolet circular dichroism of muscle proteins. Biochim. et Biophys. Acta 168, 164—167 (1968).

ORIEL, P. J.: Optical rotatory dispersion of calf thymus deoxyribonucleoprotein. Arch. Biochem. Biophys. 115, 577—582 (1966).

ORIEL, P. J., KOENIG, J. A.: The optical rotatory dispersion of MS2 bacteriophage. Arch. Biochem. Biophys. 127, 274—282 (1968).

ORIOL, C., LANDON, M. F.: Le dichroisme circulaire de diverses phosphogène phosphotrans-férases. Biochim. et Biophys. Acta 214, 455—462 (1970).

OSHIRO, Y., EYLAR, E. H.: Physical and chemical studies on glycoproteins. IV. The influence of sialic acid on the conformation of fetuin. Arch. Biochem. Biophys. 130, 227—234 (1969).

PANTALONI, A., D'ALBIS, A., DESSEN, P.: Structure et changement de conformation des pro-teines globulaires par dispersion rotatoire. J. chim. phys. 65, 196—205 (1968).

PAPKOFF, H., GOSPODAROWICZ, D., CANDIOTTI, A., LI, C. H.: Preparation of ovine interstitial cell-stimulating hormone in high yield. Arch. Biochem. Biophys. 111, 431—438 (1965).

PAPKOFF, H., SAIRAM, M. R., LI, C. H.: Amino acid sequence of the subunits of ovine pituitary interstitial cell stimulating hormone. J. Am. Chem. Soc. 93, 1531—1532 (1971).

PASTEUR, L. (1860), quoted from LOWRY, T. M.: Optical rotatory power, p. 27 (1935). New York: Dover Publ. (reprinted) 1964.

PAULING, L.: The nature of the chemical bond and the structure of molecules and crystals. An introduction to modern structural chemistry, 3rd ed. Ithaca-New York: Cornell University Press 1960.

PAULING, L., COREY, R. B.: Atomic coordinates and structure factors for two helical con-figurations of polypeptide chains. Proc. Nat. Acad. Sci. US 37, 235—240 (1951).

PAULING, L., COREY, R. B.: Stable configurations of polypeptide chains. Proc. Roy. Soc. Lon-don B. 141, 21—33 (1953).

PAULING, L., COREY, R. B.: The configuration of polypeptide chains in proteins. Fortschr. Chem. Org. Naturstoffe 11, 180—239 (1954).

PAULING, L., COREY, R. B., BRANSON, H. R.: Two hydrogen bonded helical configurations of the polypeptide chain. Proc. Nat. Acad. Sci. US 37, 205—211 (1951).

PERLMANN, G. E.: Effect of solvents and temperature on the optical rotatory properties of pepsin. Proc. Nat. Acad. Sci. US 45, 915—922 (1959).

PERLMANN, G. E., GRIZZUTI, K.: Conformational transitions of the phosphoprotein phosvitin. Random conformation → β structure. Biochemistry 10, 258—264 (1971).

PERNOLLET, J. C., GARNIER, J.: Structural modifications involved in the dissociation and re-association of the α and β subunits of ovine luteinizing hormone. FEBS Letters 18, 189—192 (1971).

PERRIN, J. H., HART, P. A.: Small molecule-macromolecule interactions as studied by optical rotatory dispersion and circular dichroism. J. Pharmac. Sci. 59, 431—448 (1970).

PFLUMM, M. N., BEYCHOK, S.: Optical activity of cystine-containing proteins. II. Circular dichroism spectra of pancreatic ribonuclease A, ribonuclease S, and ribonuclease S-protein. J. Biol. Chem. 244, 3973—3981 (1969).

PHILLIPS, D. C.: Protein crystallography in Cambridge and London: In: Aspects of protein structure (ed. G. N. RAMACHANDRAN), pp. 57—67. London-New York: Academic Press 1963.

PHILLIPS, D. C.: The three-dimensional structure of an enzyme molecule. Sci. American 215, 78—90 (1966).

PHILLIPS, D. C.: Lysozyme and the development of protein crystal chemistry. In: Proceedings of the Plenary Sessions, 7th Internat. Congress of Biochemistry, Tokyo, 1967, pp. 63—82 (1968).

PHILLIPS, D. M. P. (ed.): Histones and nucleohistones. London-New York: Plenum Press 1971.

PONSTINGL, H., HESS, M., HILSCHMANN, N.: Die vollständige Aminosäure-Sequenz des Bence-Jones-Proteins KERN. Eine neue Untergruppe der Immunoglobulin-L-Ketten vom λ-Typ. Hoppe-Seyler's Z. physiol. Chem. 349, 867—871 (1968).

PUTNAM, F. W., TITANI, K., WIKLER, M., SHINODA, T.: Structure and evolution of kappa and lambda light chains. Cold Spring Harbor Symposia Quant. Biol. 32, 9—28 (1967).

PYSH, E. S.: The calculated ultraviolet optical properties of polypeptide β-configurations. Proc. Nat. Acad. Sci. US 56, 825—832 (1966).

QUADRIFOGLIO, F., URRY, D. W.: (1) Ultraviolet rotatory properties of polypeptides in solution. I. Helical poly-L-alanine. J. Am. Chem. Soc. 90, 2755—2760 (1968).

QUADRIFOGLIO, F., URRY, D. W.: (2) Ultraviolet rotatory properties of polypeptides in solution. II. Poly-L-serine. J. Am. Chem. Soc. 90, 2760—2765 (1968).

RAMACHANDRAN, G. N., Ed.: (1) Conformation of biopolymers, vol. I. London-New York: Academic Press 1967.

RAMACHANDRAN, G. N., SASISEKHARAN, V.: Conformation of polypeptides and proteins. Advances in Protein Chem. 23, 283—437 (1968).

RAMACHANDRAN, G. N., RAMAKRISHNAN, C., SASISEKHARAN, V.: Stereochemistry of polypeptide chain configuration. J. Mol. Biol. 7, 95—99 (1963).

RAO, V. S. R., FOSTER, J. F.: Rotatory dispersion behavior of carbohydrates; a Cotton effect in D-glucose, D-xylose, and various related saccharides. Nature 200, 570—571 (1963).

RAVAL, D. N., SCHELLMAN, J. A.: The activation of chymotrypsinogen. Optical rotation studies. Biochim. et Biophys. Acta 107, 463—470 (1965).

REEKE, G. N., HARTSUCK, J. A., LUDWIG, M. L., QUIOCHO, F. A., STEITZ, T. A., LIPSCOMB, W. N.: The structure of carboxypeptidase. A. VI. Some results at 2.0 Å resolution, and the complex with glycyl-tyrosine at 2.8 Å resolution. Proc. Nat. Acad. Sci. US 58, 2220—2226 (1967).

RICHARDS, F. M., VITHAYATHIL, P. J.: The preparation of subtilisin-modified ribonuclease and the separation of the peptide and protein components. J. Biol. Chem. 234, 1459—1465 (1959).

RIDDIFORD, L. M.: Solvent perturbation and ultraviolet optical rotatory dispersion studies of paramyosin. J. Biol. Chem. 241, 2792—2802 (1966).

RIDGEWAY, D.: Polarimetric analysis of protein structure. Advances in Biol. and Med. Phys. 9, 271—353 (1963).

ROSENFELD, L.: Quantenmechanische Theorie der natürlichen optischen Aktivität von Flüssigkeiten und Gasen. Z. Physik 52, 161—174 (1928).

ROSENHECK, K., DOTY, P.: The far ultraviolet absorption spectra of polypeptide and protein solutions and their dependence on conformation. Proc. Nat. Acad. Sci. US 47, 1175—1185 (1961).

ROSENHECK, K., SOMMER, B.: Theory of the far-ultraviolet sepctrum of polypeptides in the β conformation. J. Chem. Phys. 46, 532—537 (1967).

ROSENKRANZ, H., SCHOLTAN, W.: Cirkulardichroismus und Konformation der L-Asparaginase. H.-S. Z. physiol. Chem. 352, 1081—1090 (1971).

ROSENKRANZ, H., SCHOLTAN, W.: Eine verbesserte Methode zu Konformationsbestimmung von helicalen Proteinen aus Messungen des Cirkulardichroismus. H.-S. Z. physiol. Chem. 352, 896—904 (1971).

ROSS, D. L.: Conformational studies of immunoglobulins IgG and antigen-antibody complexes. Dissertation. Houston: University of Texas Grad. School of Biomed. Science 1967.

Ross, D. L., Jirgensons, B.: The far ultraviolet optical rotatory dispersion, circular dichroism, and absorption spectra of a myeloma immunoglobulin IgG. J. Biol. Chem. **243**, 2829—2836 (1968).

Rossmann, M. G., Jeffrey, B. A., Main, P., Warren, S.: The crystal structure of lactic dehydrogenase. Proc. Nat. Acad. Sci. US **57**, 515—524 (1967).

Rossmann, M. G., Adams, M. J., Buehner, M., Ford, G. C., Hackert, M. L., Lentz, P. J., McPherson, A., Schewitz, R. W., Smiley, I. E.: Structural constraints of possible mechanisms of lactate dehydrogenase as shown by high resolution studies of the apoenzyme and a variety of enzyme complexes. Cold Spring Harbor Symposia Quant. Biol. **36**, 179—191 (1972).

Rudolph, H.: Photoelectric polarimeter attachment. J. Opt. Soc. Am. **45**, 50—59 (1955).

Rupley, J. A.: The comparison of protein structure in the crystal and in solution. In: Structure and stability of biological macromolecules (eds. S. N. Timasheff, G. D. Fasman), pp. 291—352. New York: M. Dekker, Inc. 1969.

Ruttenberg, M. A., King, T. P., Craig, L. C.: The use of the tyrocidines for the study of conformation and aggregation behavior. J. Am. Chem. Soc. **87**, 4196—4198 (1965).

Samejima, T., Yang, J. T.: Optical rotatory dispersion of DNA and RNA. Biochemistry **3**, 613—616 (1964).

Sarin, P. S., Zamecnik, P. C., Bergquist, P. L., Scott, J. F.: Conformational differences among purified samples of transfer RNA from yeast. Proc. Nat. Acad. Sci. US **55**, 579—585 (1966).

Sarkar, P. K., Doty, P.: The optical rotatory properties of the β-configuration in polypeptides and proteins. Proc. Nat. Acad. Sci. US **55**, 981—989 (1969).

Sarkar, P. K., Yang, J. T., Doty, P.: Optical rotatory dispersion of E. coli ribosomes and their constituents. Biopolymers **5**, 1—4 (1967).

Sauer, K., Dratz, E. A., Coyne, L.: Circular dichroism spectra and the molecular arrangement ob bacteriochlorophylls in the reaction centers of photosynthetic bacteria. Proc. Nat. Acad. Sci. US **61**, 17—24 (1968).

Scanu, A.: Studies on the conformation of human serum high-density lipoproteins HDL_2 and HDL_3. Proc. Nat. Acad. Sci. US **54**, 1699—1705 (1965).

Scanu, A., Hirz, R.: Human serum low-density lipoprotein protein: its conformation studied by circular dichroism. Nature **218**, 200—201 (1968).

Scanu, A., Hirz, R.: On the structure of human serum high-density lipoprotein: studies by the technique of circular dichroism. Proc. Nat. Acad. Sci. US **59**, 890—894 (1968).

Scanu, A. M.: On the temperature dependence of the conformation of human serum high density lipoprotein. Biochim. et Biophys. Acta **181**, 268—274 (1969).

Scanu, A. M., Pollard, H., Hirz, R., Kothary, K.: On the conformational instability of human serum low-density lipoprotein: effect of temperature. Proc. Nat. Acad. Sci. US **62**, 171—178 (1969).

Scanu, A. M., van Deenen, L. L. M., de Haas, G. H.: Optical rotatory dispersion and circular dichroism of phospholipase A2 and its zymogen from porcine pancreas. Biochim. et Biophys. Acta **181**, 471—473 (1969).

Schechter, I., Berger, A.: Peptides of L- and D-alanine. Synthesis and optical rotations. Biochemistry **5**, 3362—3370 (1966).

Schechter, B., Schechter, I., Ramachandran, J., Conway-Jacobs, A., Sela, M.: The synthesis and circular dichroism of a series of peptides possessing the structure (L-tyrosyl-L-alanyl-L-glutamyl)$_n$. Eur. J. Biochem. **20**, 301—308 (1971).

Schellman, C. G., Schellman, J. A.: The optical rotatory power of proteins and polypeptides. VII. Optical rotation and protein configuration. Compt. rend. trav. lab. Carlsberg. Ser. chim. **30**, 463—499 (1958).

Schellman, J. A.: The optical rotatory properties of proteins and polypeptides. I. Theoretical discussion and experimental methods. Compt. rend. trav. lab. Carlsberg. Ser. chim. **30**, 363—394 (1958).

Schellman, J. A., Oriel, P.: Origin of the Cotton effect of helical polypeptides. J. Chem. Phys. **37**, 2114—2124 (1962).

SCHELLMAN, J. A.: The rotatory properties of natural amino acids. In: Optical rotatory dispersion, applications to organic chemistry (ed. C. DJERASSI), pp. 210—228. New York-Toronto-London: McGraw-Hill Co. 1960.

SCHELLMAN, J. A., LOWE, M. J.: The optical rotation of ribonuclease. J. Am. Chem. Soc. 90, 1070—1072 (1968).

SCHELLMAN, J. A., SCHELLMAN, C. G.: Use of rotatory dispersion in the determination of protein structure. J. Polymer Sci. 49, 129—151 (1961).

SCHELLMAN, J. A., SCHELLMAN, C. G.: The conformation of polypeptide chains in proteins. In: The proteins, composition, structure, and function (ed. H. NEURATH), 2nd ed., vol. II, pp. 1—137. New York-London: Academic Press 1964.

SCHERAGA, H. A.: Intramolecular bonds in proteins. II. Noncovalent bonds. In: The proteins, composition, structure, and function (ed. H. NEURATH), 2nd ed., vol. I, pp. 477—594. New York-London: Academic Press 1963.

SCHERAGA, H. A., SCOTT, R. A., VANDERKOOI, G., LEACH, S. J., GIBSON, K. D., OOI, T., NEMETHY, G.: Calculations of polypeptide structures from amino acid sequence. In: Conformation of biopolymers (ed. G. N. RAMACHANDRAN), pp. 43—60. London-New York: Academic Press 1967.

SCHNEIDER, A. S., SCHNEIDER, M. J. T., ROSENHECK, K.: Optical activity of biological membranes: scattering effects and protein conformation. Proc. Nat. Acad. Sci. US 66, 793—798 (1970).

SCHOLTAN, W., ROSENKRANZ, H.: Optisches Drehungsvermögen, Zirkulardichroismus und Ultraviolettspektrum des Kallikrein-Inaktivators. Makromol. Chem. 99, 254—274 (1966).

SCHULTZE, H. E., HEREMANS, J. F.: Molecular biology of human proteins with special reference to plasma proteins, vol. I: Nature and metabolism of extracellular proteins. Amsterdam-London-New York: Elsevier Publ. 1966.

SEIFTER, S., GALLOP, P. M.: The structure proteins. In: The proteins, composition, structure, and function (ed. H. NEURATH), 2nd ed., vol. IV, pp. 155—458. New York-London: Academic Press 1966.

SELA, M., STEINBERG, I. Z., DANIEL, E.: Optical rotatory properties of poly-L-tryptophan. Biochim. et Biophys. Acta 46, 433—440 (1961).

SHECHTER, E., BLOUT, E. R.: An analysis of the optical rotatory dispersion of polypeptides and proteins. Proc. Nat. Acad. Sci. US 51, 695—702 (1964).

SHIBATA, Y., KRONMAN, M. J.: Inactivation of glyceraldehyde-3-phosphate dehydrogenase. Arch. Biochem. Biophys. 118, 410—419 (1967).

SHIH, T. Y., FASMAN, G. D.: Circular dichroism studies on DNA complexes with arginine-rich histone IV (f2a1). Biochemistry 10, 1675—1683 (1971).

SHICHI, H., LEWIS, M. S., IRREVERRE, F., STONE, A. L.: Biochemistry of visual pigments. I. Purification and properties of bovine rhodopsin. J. Biol. Chem. 244, 529—536 (1969).

SIMMONS, N. S., BLOUT, E. R.: The structure of tobacco mosaic virus and its components. Biophys. J. 1, 55—62 (1960).

SIMMONS, N. S., COHEN, C., SZENT-GYORGYI, A. G., WETLAUFER, D. B., BLOUT, E. R.: A conformation-dependent Cotton effect in alpha-helical polypeptides and proteins. J. Am. Chem. Soc. 83, 4766—4769 (1961).

SIMONS, E. R., FASMAN, G. D., BLOUT, E. R.: The pepsin-catalyzed hydrolysis of the α-helical form of poly-α-L-glutamic acid. J. Biol. Chem. 236, 64—65 (1961).

SIMPSON, R. T., VALLEE, B. L.: Side chain Cotton effects of ribonuclease. Biochemistry 5, 2531—2538 (1966).

SIMPSON, R. T., SOBER, H. A.: Circular dichroism of calf liver nucleohistone. Biochemistry 9, 3103—3109 (1970).

SINGHAL, R. P., ATASSI, M. Z.: Conformational studies on modified proteins and peptides. Conformation of peptides with intact and overlapping helices obtained by cleavage of myoglobin at proline peptide bonds. Biochemistry 9, 4252—4259 (1970).

SMILLIE, L. B., ENENKEL, A. G., KAY, C. M.: Physicochemical properties and amino acid composition of chymotrypsinogen B. J. Biol. Chem. 241, 2097—2102 (1966).

SNATZKE, G. (ed.): Optical rotatory dispersion and circular dichroism in organic chemistry. London: Heyden and Son 1967.

SNATZKE, G.: Circular dichroism and optical rotatory dispersion. Principles and applications to the investigation of the stereochemistry of natural products. Angew. Chem. Internat. Ed. 7, 14—26 (1968).

SOBER, H. A. (ed.): Handbook of biochemistry, Selected data for molecular biology, 2nd ed. Cleveland, Ohio: The Chemical Rubber Publ. 1970.

SONENBERG, M., BEYCHOK, S.: Circular dichroism studies of biologically active growth hormone preparations. Biochim. Biophys. Acta 229, 88—101 (1971).

SORENSEN, M. E., HERSKOVITS, T. T.: Optical activity and solvent perturbation of fumarase. Biochim. et Biophys. Acta 257, 20—29 (1972).

SPIRIN, A. S., GAVRILOVA, L. P.: The ribosomes. In: Molecular biology, biochemistry and biophysics (eds. W. KLEINZELLER, G. F. SPRINGER, H. G. WITTMANN), vol. 4. Berlin-Heidelberg-New York: Springer 1969.

SPRINGER, G. F.: Human MN glycoproteins: dependence of blood-group and anti-influenza virus activities on their molecular size. Biochem. Biophys. Res. Communs. 28, 510—513 (1967).

SPRINGER, G. F., DESAI, P. R., KOLECKI, B.: Synthesis and immunochemistry of fucose methylethers and their methylglycosides. Biochemistry 3, 1076—1085 (1964).

SPRINGER, G. F., NAGAI, Y., TEGTMEYER, H.: Isolation and properties of human blood-group NN and meconium-Vg antigens. Biochemistry 5, 3254—3272 (1966).

SPRINGER, G. F., DESAI, P. R.: Monosaccharides as specific precipitinogens of eel anti-human blood-group H(0) antibody. Biochemistry 10, 3749—3761 (1971).

SPRINGER, G. F., ADYE, J. C., BEZKOROVAINY, A., MURTHY, J. R.: Functional aspects and nature of the lipopolysaccharide-receptor of human erythrocytes (in press).

SRERE, P. A., BROOKS, G. C.: The circular dichroism of glucagon solutions. Arch. Biochem. Biophys. 129, 708—710 (1969).

STAPRANS, I., WATANABE, S.: Optical properties of troponin, tropomyosin, and relaxing protein of rabbit skeletal muscle. J. Biol. Chem. 245, 5962—5966 (1970).

STAUDINGER, H.: Über die Einteilung der Kolloide. Ber. deutsch. chem. Ges. 68, 1682—1691 (1935).

STAUDINGER, H.: Organische Kolloidchemie, 3. Aufl. Braunschweig: F. Vieweg & Sohn 1950.

STAUDINGER, H.: Die hochmolekularen organischen Verbindungen — Kautschuk und Cellulose. Berlin: Springer 1932. Ann Arbor, Michigan: Edwards Brothers Inc., Publ. (reprinted) 1943.

STAUDINGER, H.: Arbeitserinnerungen. Heidelberg: Dr. A. Huethig Verlag GmbH 1961.

STAUDINGER, H., FRITSCHI, J.: Über die Hydrierung des Kautschuks und über seine Konstitution. Helv. Chim. Acta 5, 785—806 (1922).

STEINER, L. A., LOWEY, S.: Optical rotatory dispersion studies of rabbit γ-G-immunoglobulin and its papain fragments. J. Biol. Chem. 241, 231—240 (1966).

STEINER, L. A., PORTER, R. R.: The interchain disulfide bonds of a human pathological immunoglobulin. Biochemistry 6, 3957—3970 (1967).

STONE, A. L.: Optical rotatory dispersion of mucopolysaccharides and mucopolysaccharide-dye complex. Biopolymers 3, 617—624 (1965).

STONE, A. L.: Conformation of hexose polysaccharides in solution. In: Structure and stability of biological macromolecules (eds. S. N. TIMASHEFF, G. D. FASMAN), pp. 353—413. New York: M. Dekker Inc. 1969.

STONE, A. L.: Optical rotatory dispersion of mucopolysaccharides. III. Ultraviolet circular dichroism and conformational specificity in amide groups. Biopolymers 10, 739—751 (1971).

STRAUB, F. B.: SH groups and SS bridges in the structure of enzymes. In: Proceeding of the Plenary Sessions, 7th Internat. Congress of Biochemistry, I. U. B., Tokyo, Aug. 20—25, 1967; vol. 36, 41—50 (1968).

STRAUSS, J. H., GORDON, A. S., WALLACH, D. F. H.: The influence of tertiary structure upon the optical activity of three globular proteins: myoglobin, hemoglobin, and lysozyme. Eur. J. Biochem. 11, 201—212 (1969).

STRICKLAND, E. H., HARDIN, E.: Circular dichroism of horseradish peroxidase and its enzyme-substrate compounds. Biochim. et Biophys. Acta 151, 70—75 (1968).

STRICKLAND, E. H., KAY, E., SHANNON, L. M., HORWITZ, J.: Peroxidase isoenzymes from horseradish roots. III. Circular dichroism of isoenzymes and apoisoenzymes. J. Biol. Chem. 243, 3560—3565 (1968).

STRICKLAND, E. H., HORWITZ, J., BILLUPS, C.: Fine structure in the near-ultraviolet circular dichroism and absorption spectra of tryptophan derivatives and chymotrypsinogen A at 77° K. Biochemistry 8, 3205—3213 (1969).

STRICKLAND, E. H., KAY, E., SHANNON, L. M.: Effects of denaturing agents on the phenylalanyl circular dichroism bands of horseradish peroxidase isoenzymes and apoisoenzymes. J. Biol. Chem. 245, 1233—1238 (1970).

STRICKLAND, E. H., WILCHEK, M., HORWITZ, J. BILLUPS, C.: Low temperature circular dichroism of tyrosyl and tryptophanyl diketopiperazines. J. Biol. Chem. 245, 4168—4177 (1970).

STROUD, R. M., KAY, L. M., DICKERSON, R. E.: The crystal and molecular structure of DIP-inhibited bovine trypsin at 2.7 Å resolution. Cold Spring Harbor Symposia Quant. Biol. 36, 125—140 (1972).

STURTEVANT, J. M., TSONG, T. Y.: Investigation of yeast L-lactate dehydrogenase (cytochrome b₂). VI. Circular dichroism of the holoenzyme. J. Biol. Chem. 244, 4942—4950 (1969).

SUGITA, Y., DOHI, Y., YONEYAMA, Y.: Circular dichroism of lamprey and human hemoglobins. Biochem. Biophys. Res. Communs. 31, 447—452 (1968).

SUND, H., WEBER, K.: The quaternary structure of proteins. Angew. Chem. Internat. Ed. 5, 231—245 (1966).

SVEDBERG, T., PEDERSEN, K. O.: The ultracentrifuge. Oxford: Clarendon Press 1940.

TAMBURRO, A. M., SCATTURIN, A., MORODER, L.: Far-ultraviolet optical rotatory dispersion and circular dichroism studies of bovine pancreatic ribonuclease A. Biochim. et Biophys. Acta 154, 583—585 (1968).

TAMBURRO, A. M., SCATTURIN, A., ROCCHI, R.: A conformational study on pepsin and trypsin. Gazz. chim. ital. 98, 1256—1260 (1968).

TAMBURRO, A. M., SCATTURIN, A., GRAZI, E., PONTREMOLI, S.: Conformational states of rabbit liver fructose 1,6-diphosphatase. J. Biol. Chem. 245, 6624—6627 (1970).

TAMBURRO, A. M., SCATTURIN, A., VIDALI, G.: On the conformation of histones: circular dichroism studies on the lysine- and serine-rich fractions from avian erythrocytes. Intern. J. Protein Res. 2, 127—131 (1970).

TABORSKY, G.: Optical rotatory dispersion and circular dichroism of phosvitin at low pH. Reversible transition between unordered conformation and β-structure. J. Biol. Chem. 243, 6014—6020 (1968).

TAKESADA, H., HAMAGUCHI, K.: Circular dichroism of hemocyanin. J. Biochem. (Tokyo) 63, 725—729 (1968).

TANFORD, C.: Physical chemistry of macromolecules. New York-London: J. Wiley & Sons 1961.

TANFORD, C.: The effect of amino acid composition on the optical rotatory dispersion of randomly coiled proteins. Abstracts, 7th Internat. Congress of Biochem., Tokyo, Symp. I 3, 5, pp. 33—34 (1967).

TANFORD, C.: Protein denaturation. Advan. Protein Chem. 23, 121—282 (1968).

TANFORD, C., DE, P. K., TAGGART, V. G.: The role of the α-helix in the structure of proteins. Optical rotatory dispersion of β-lactoglobulin. J. Am. Chem. Soc. 82, 6028—6034 (1960).

TANFORD, C., KAWAHARA, K., LAPANJE, S.: Proteins as random coils. I. Intrinsic viscosities and sedimentation coefficients in concentrated guanidine hydrochloride. J. Am. Chem. Soc. 89, 729—736 (1967).

TANFORD, C., KAWAHARA, K., LAPANJE, S., HOOKER, T. M., ZARLENGO, M. H., SALAHUDDIN, A., AUNE, K. C., TAKAGI, T.: Proteins as random coils. III. Optical rotatory dispersion in 6 M guanidine hydrochloride. J. Am. Chem. Soc. 89, 5023—5029 (1967).

TANG, S. P. W., COLEMAN, J. E.: Optical rotatory dispersion of copper proteins. Pseudomonas blue protein. Biochem. Biophys. Res. Communs. 27, 281—286 (1967).

TANG, S.-P. W., COLEMAN, J. E., MYER, Y. P.: Conformational studies of copper proteins. J. Biol. Chem. 243, 4286—4297 (1968).

TANIUCHI, H., ANFINSEN, C. B.: Steps in the formation of active derivatives of Staphylococcal nuclease during trypsin digestion. J. Biol. Chem. 243, 4778—4786 (1968).

TEICHBERG, V. I., KAY, C. M., SHARON, N.: Separation of contributions of tryptophans and tyrosines to the ultraviolet circular dichroism spectrum of hen egg-white lysozyme. Eur. J. Biochem. 16, 55—59 (1970).

TERMINE, J. D., EANES, E. D., EIN, D., GLENNER, G. G.: Infrared spectroscopy of human amyloid fibrils and immunoglobulin proteins. Biopolymers 11, 1103—1113 (1972).

THORNE, C. J. R., KAPLAN, N. O.: Physicochemical properties of pig and horse heart mitochondrial malate dehydrogenase. J. Biol. Chem. 238, 1861—1868 (1963).

TIFFANY, M. L., KRIMM, S.: Circular dichroism of the "random" polypeptide chain. Biopolymers 8, 347—359 (1969).

TIMASHEFF, S. N., SUSI, H.: Infrared investigation of the secondary structure of β-lactoglobulins. J. Biol. Chem. 241, 249—251 (1966).

TIMASHEFF, S. N., SUSI, H., TOWNEND, R., STEVENS, L., GORBUNOFF, M. J., KUMOSINSKI, T. F.: Application of circular dichroism and infrared spectroscopy to the conformation of proteins in solution. In: Conformation of biopolymers (ed. G. N. RAMACHANDRAN), vol. 1, pp. 173—196. London-New York: Academic Press 1967.

TIMASHEFF, F. N., TOWNEND, R.: The circular dichroism of phosvitin. J. Biol. Chem. 242, 2290—2292 (1967).

TIMASHEFF, F. N., TOWNEND, R., MESCANTI, L.: The optical rotatory dispersion of the β-lactoglobulins. J. Biol. Chem. 241, 1863—1870 (1966).

TIMASHEFF, S. N., BERNARDI, G.: Studies on acid deoxyribonuclease. VII. Conformation of three nucleases in solution. Arch. Biochem. Biophys. 141, 53—58 (1970).

TINOCO, I., JR.: Absorption and rotation of polarized light by polymers. In: Molecular biophysics (eds. B. PULLMAN, M. WEISSBLUTH), pp. 269—291. New York-London: Academic Press 1965.

TINOCO, I., CANTOR, C. R.: Application of optical rotatory dispersion and circular dichroism to the study of biopolymers. Meth. biochem. Anal. 18, 81—203 (1970).

TODD, A.: Optical rotation. In: A laboratory manual of analytical methods of protein chemistry (eds. P. ALEXANDER, R. J. BLOCK), vol. II, pp. 245—283. New York-Oxford-London-Paris: Pergamon Press 1960.

TOMIMATSU, Y., GAFFIELD, W.: Optical rotatory dispersion of egg proteins. I. Ovalbumin, conalbumin, ovomucoid, and lysozyme. Biopolymers 3, 509—517 (1965).

TOMIMATSU, Y., VITELLO, L., GAFFIELD, W.: Effect of aggregation on the optical rotatory dispersion of poly(α, L-glutamic acid). Biopolymers 4, 653—662 (1966).

TORCHINSKII, Y. M., KORENEVA, L. G.: Effect of substrate analogs and carbonyl group reagents on the anomalous dispersion of aspartate-glutamate transaminase. Biokhimiya 29, 780—790 (1964).

TOWNEND, R., KUMOSINSKI, T. F., TIMASHEFF, S. N.: The circular dichroism of variants of β-lactoglobulin. J. Biol. Chem. 242, 4538—4545 (1967).

TRAYER, H. R., BUCKLEY, C. E.: Molecular properties of lysostaphin, a bacteriolytic agent specific for *Staphylococcus aureus*. J. Biol. Chem. 245, 4842—4846 (1970).

TROITSKII, G. V.: Application of an extended Moffitt equation for the evaluation of protein conformation. A proof of the presence of the β structures in many globular proteins. Biofizika 10, 895—901 (1965).

TSONG, T. Y., STRURTEVANT, J. M.: Investigation of yeast L-lactate dehydrogenase (cytochrome b₂). V. Circular dichroism of the flavin mononucleotide-free apoenzyme. J. Biol. Chem. 244, 2397—2402 (1969).

TUAN, D. Y. H., BONNER, J.: Optical absorbance and rotatory dispersion studies on calf thymus nucleohistone. J. Mol. Biol. 45, 59—76 (1969).

ULMER, D. D.: Optical rotatory dispersion of a heme peptide from cytochrome c. Proc. Nat. Acad. Sci. US 55, 894—899 (1966).

ULMER, D. D., VALLEE, B. L.: Extrinsic Cotton effects and the mechanism of enzyme action. Advances in Enzymol. 27, 37—104 (1965).

URNES, P.: The crystal-solution problem of sperm whale myoglobin. Thesis. Cambridge, Massachusetts: Harvard University 1963.

URNES, P., DOTY, P.: Optical rotatory dispersion and the conformation of polypeptides and proteins. Advan. Protein Chem. **16**, 401—544 (1961).

URRY, D. W.: Protein-heme interactions in heme proteins: cytochrome c. Proc. Nat. Acad. Sci. US **54**, 640—648 (1965).

URRY, D. W., EYRING, H.: The role of the physical sciences in biomedical research. Perspectives in Biology and Medicine **1966**, 450—475.

URRY, D. W., MEDNIEKS, M., BEJNAROWICZ, E.: Optical rotation of mitochondrial membranes. Proc. Nat. Acad. Sci. US **57**, 1043—1049 (1967).

URRY, D. W.: Circular dichroism pattern of methylpyrrolidone can resemble that of the α-helix. J. Phys. Chem. **72**, 3035—3038 (1968).

URRY, D. W., JI, T. H.: Distortions in circular dichroism patterns of particulate (or membraneous) systems. Arch. Biochem. Biophys. **128**, 802—807 (1968).

URRY, D. W., QUADRIFOGLIO, F., WALTER, R., SCHWARTZ, I. L.: Conformational studies on neurohypophyseal hormones; the disulfide bridge of oxytocin. Proc. Nat. Acad. Sci. US **60**, 967—974 (1968).

URRY, D. W.: Optical rotation and biomolecular conformation. In: Spectroscopic approaches to biomolecular conformation (ed. D. W. URRY), pp. 33—121. Chicago: Am. Med. Assoc. 1970.

URRY, D. W., MASOTTI, L., KRIVACIC, J. R.: Circular dichroism of biological membranes. I. Mitochondria and red blood cell ghosts. Biochim. et Biophys. Acta **241**, 600—612 (1971).

VAN'T HOFF, J. H.: A treatise on a system of atomic formulae in three dimensions and on the relation between rotatory power and chemical constitution. Pamphlet in Dutch, 1874; French transl., 1875. See also: J. H. VAN'T HOFF, Imagination in science (transl. by G. F. SPRINGER). In: Molecular biology biochemistry and biophysics (eds. W. KLEINZELLER, G. F. SPRINGER, H. G. WITTMANN), vol. 1. Berlin-Heidelberg-New York: Springer 1967.

VELLUZ, L., LEGRAND, M., GROSJEAN M.: Optical circular dichroism. Principles, measurements, and applications. Weinheim: Verlag Chemie; New York-London: Academic Press 1965.

VENKATACHALAM, C. M., RAMACHANDRAN, G. N.: Stereochemistry of polypeptide chains — comparison of different potential functions. In: Conformation of biopolymers (ed. G. N. RAMACHANDRAN), pp. 83—105. London-New York: Academic Press 1967.

VENTILLA, M., CANTOR, C. R., SHELANSKI, M.: A circular dichroism study of microtubule protein. Biochemistry **11**, 1554—1561 (1972).

VERDET, M. (1854), quoted from T. M. LOWRY: Optical rotatory power, p. 162 (1935). New York: Dover Publ. (reprinted) 1964.

VERPOORTE, J. A., KAY, C. M.: Optical rotatory dispersion and spectrophotometric studies on fetuin in solutions containing detergent and denaturing reagents. Biochim. et Biophys. Acta **126**, 551—569 (1966).

VINOGRADOV, S., ZAND, R.: Circular dichroism studies. I. Cytochrome c. Arch. Biochem. Biophys. **125**, 902—910 (1968).

VISSER, L., BLOUT, E. R.: Elastase. II. Optical properties and the effect of sodium dodecyl sulfate. Biochemistry **10**, 743—752 (1971).

VOURNAKIS, J. N., YAN, J. F., SCHERAGA, H. A.: Effect of side chains on the conformational energy and rotational strength of the n-π^* transition for some α-helical poly-α-amino acids. Biopolymers **6**, 1536—1550 (1968).

WAKS, M., ALFSEN, A.: Structural studies of haptoglobins. Hydrogen ion equilibria and optical rotatory dispersion of Hp 1—1 and Hp 2—2. Arch. Biochem. Biophys. **113**, 304—313 (1966).

WALBORG, E. F., WARD, D. N.: The carbohydrate components of ovine luteinizing hormone. Biochim. et Biophys. Acta **78**, 304—312 (1963).

WARD, D. N.: Gonadotropic hormones. In: The encyclopedia of biochemistry (eds. R. J. WILLIAMS, E. M. LANSFORD), pp. 384—386. New York-Amsterdam-London: Reinhold Publ. 1967.

WARD, D. N., WALBORG, E. F., ADAMS-MAYNE, M.: Amino acid composition and electrophoretic properties of ovine luteinizing hormone. Biochim. et Biophys. Acta **50**, 224—232 (1961).

WARD, D. N., ADAMS-MAYNE, M., RAY, N., BALKE, D. E., COFFEY, J., SHOWALTER, M.: Comparative studies of luteinizing hormone from beef, pork, and sheep pituitaries. I. Purification and physical properties. General and Comparat. Endocrinology 8, 44—53 (1967).

WELLNER, D.: Evidence for conformational changes in L-amino- acid oxidase associated with reversible inactivation. Biochemistry 5, 1585—1591 (1966).

WETTER, O., JAHNKE, K., HERTENSTEIN, C.: Die optische Rotationsdispersion isolierter Paraproteine. II. Zur Gruppenspezifität pathologischer γ-M-Proteine. Klin. Wochenschr. 44, 573—579 (1966).

WHITE, F. H., JR.: Regeneration of enzymic activity by air-oxidation of reduced ribonuclease with observations on thiolation during reduction with thioglycolate. J. Biol. Chem. 235, 383—389 (1961).

WHITNEY, P. L., TANFORD, C.: Recovery of specific activity after complete unfolding and reduction of an antibody fragment. Proc. Nat. Acad. Sci. US 53, 524—532 (1965).

WILKINS, M. H. F., ZUBAY, G., WILSON, H. R.: X-ray diffraction studies of the molecular structure of nucleohistones and chromosomes. J. Mol. Biol. 1, 179—185 (1959).

WILSON, E. M., MEISTER, A.: Optical rotatory dispersion of L-aspartate β-decarboxylase and its derivatives. Biochemistry 5, 1166—1174 (1966).

WILSON, W. D., FOSTER, J. F.: Photolysis accompanying peptide absorption in proteins. Biophysical J. 12, 609—624 (1972).

WOLDBYE, F.: Instrumentation. In: Optical rotatory dispersion and circular dichroism in organic chemistry (ed. G. SNATZKE), pp. 85—100. London: Heyden and Son 1967.

WOLFE, F. H.: Ultraviolet circular dichroism of wheat embrio ribosomal ribonucleates. Biochemistry 7, 3361—3366 (1968).

WOLFE, F. H., OIKAWA, K., KAY, C. M.: Optical rotatory dispersion and absorbance temperature studies on wheat embrio ribosomal ribonucleates. Can. J. Biochem. and Physiol. 46, 643—653 (1968).

WOODY, R. W.: Optical properties of polypeptides in the β-conformation. Biopolymers 8, 669—683 (1969).

WOODY, R. W., TINOCO, I.: Optical rotation of oriented helices. III. Calculation of the rotatory dispersion and circular dichroism of alpha- and 3_{10}-helix. J. Chem. Phys. 46, 4927—4945 (1967).

WRIGHT, C. S., ALDEN, R. A., KRAUT, J.: Structure of subtilisin BPN' at 2.5 Å resolution. Nature 221, 235—242 (1969).

WU, V., SCHERAGA, H. A.: Studies of soybean trypsin inhibitor. I. Physicochemical properties. Biochemistry 1, 698—705 (1962).

WU, J. Y., YANG, J. T.: Physicochemical characterization of citrate synthase and its subunits. J. Biol. Chem. 245, 212—218 (1970).

WU, T. T., KABAT, E. A.: An attempt to locate the non-helical and permissively helical sequences of proteins: application to the variable regions of immunoglobulin light and heavy chains. Proc. Nat. Acad. Sci. US 68, 1501—1506 (1971).

WYCKOFF, H. W., HARDMAN, K. D., ALLEWELL, N. M., INAGAMI, T., JOHNSON, L. N., RICHARDS, F. M.: The structure of ribonuclease-S at 3.5 Å resolution. J. Biol. Chem. 242, 3984—3988 (1967).

YANG, C. C., CHANG, C. C., HAMAGUCHI, K., IKEDA, K., HAYASHI, K., SUZUKI, T.: Optical rotatory dispersion of cobrotoxin. J. Biochem. (Tokyo) 61, 272—274 (1967).

YANG, J. T.: On the phenomenological treatments of optical rotatory dispersion of polypeptides and proteins. Proc. Nat. Acad. Sci. US 53, 438—445 (1965).

YANG, J. T.: (1) Optical rotatory dispersion. In Poly-α-amino acids, protein models for conformational studies (ed. G. D. FASMAN), pp. 239—291. New York: M. Dekker, Inc. 1967.

YANG, J. T.: (2) Optical activity of the α, β and coiled conformations in polypeptides and proteins. In: Conformation of biopolymers (ed. G. N. RAMACHANDRAN), vol. 1, pp. 157—172. London-New York: Academic Press 1967.

YANG, J. T., DOTY, P.: The optical rotatory dispersion of polypeptides and proteins in relation to configuration. J. Am. Chem. Soc. 79, 761—775 (1957).

YANG, J. T., SAMEJIMA, T.: Optical rotatory dispersion of catalase. J. Biol. Chem. 238, 3262—3267 (1963).

YANG, J. T., SAMEJIMA, T.: Optical rotatory dispersion and circular dichroism of nucleic acids. Prog. Nucl. Acid Res. Mol. Biol. 9, 223—300 (1969).

YOSHIDA, F., MOTAI, H., ICHISHIMA, E.: Physical and chemical properties of lipase from *Torulopsis ernorhii*. Biochim. et Biophys. Acta 154, 586—588 (1968).

ZAND, R., AGRAWAL, B. B. L., GOLDSTEIN, I. J.: pH-dependent conformational changes of concanavalin A. Proc. Nat. Acad. Sci. US 68, 2173—2176 (1971).

ZIEGLER, S. M., BUSH, C. A.: Circular dichroism of cyclic hexapeptides with one or two side chains. Biochemistry 10, 1330—1335 (1971).

ZUBAY, G., DOTY, P.: The isolation and properties of deoxyribonucleoprotein particles containing single nucleic acid molecules. J. Mol. Biol. 1, 1—20 (1959).

ZUBKOV, V. A., BIRSHTEIN, T. M., MILEVSKAYA, I. S., VOLKENSTEIN, M. V.: Circular dichroism of random coil polypeptides (in Russian). Molekul. Biol. 4, 715—723 (1970).

A Selection of Recent References Added in Proof

ADYE, J. C., SPRINGER, G. F., MURTY, J. R.: On the nature and function of the lipopolysaccharide receptor from human erythrocytes. Z. Immun.-Forsch. 144, 491—496 (1973).

ALOJ, S., EDELHOCH, H.: Structural studies on polypeptide hormones. II. Parathyroid hormone. Arch. Biochem. Biophys. 150, 782—785 (1972).

ANFINSEN, C. B.: Principles that govern the folding of polypeptide chains. Science 181, 223—230 (1973).

ATASSI, M. Z., SINGHAL, R. P.: Conformation studies on modified proteins and peptides. V. Conformation of myoglobin derivatives modified at two carboxyl groups. J. Biol. Chem. 247, 5980—5986 (1972).

AUER, H. E.: Far-ultraviolet absorption and circular dichroism spectra of L-tryptophan and some derivatives. J. Am. Chem. Soc. 95, 3003—3011 (1973).

AZUMA, M., AZUMA, K., KITO, Y.: Circular dichroism of visual pigment analogues containing 3-dehydroretinal and 5,6-epoxy-3-dehydroretinal as the chromophore. Biochim. et Biophys. Acta 295, 520—527 (1973).

AZUMA, T., HAMAGUCHI, K., MIGITA, S.: Acid denaturation of Bence Jones proteins. J. Biochem. (Tokyo) 71, 379—386 (1972).

BAREL, A. O., PRIEELS, J. P., MAES, E., LOOZE, Y., LÉONIS, J.: Comparative physicochemical studies of human α-lactalbumin and human lysozyme. Biochim. et Biophys. Acta 257, 288—296 (1972).

BAYER, E., BACHER, A., KRAUSS, P., VOELTER, W., BARTH, G., BUNNENBERG, E., DJERASSI, C.: Investigation of xanthine oxidase. Magnetic CD studies. Eur. J. Biochem. 22, 580—584 (1971).

BEWLEY, T. A., LI, C. H.: Circular dichroism studies on human pituitary growth hormone and ovine pituitary lactogenic hormone. Biochemistry 11, 884—888 (1972).

BEWLEY, T. A., LI, C. H.: Molecular weight and circular dichroism studies of bovine and ovine pituitary growth hormones. Biochemistry 11, 927—931 (1972).

BIRKTOFT, J. J., BLOW, D. M.: Structure of crystalline α-chymotrypsin. V. The atomic structure of tosyl-α-chymotrypsin at 2 A resolution. J. Mol. Biol. 68, 187—240 (1972).

BJÖRK, I., STENFLO, J.: A conformation study of normal and dicoumarol-induced prothrombin. FEBS Letters 32, 343—346 (1973).

BLAKE, C. C. F.: X-ray studies of crystalline proteins. Progr. Biophys. Mol. Biol. 25, 83—130 (1972).

BLAUER, G., HARMATZ, D., SNIR, J.: Optical properties of bilirubin-serum albumin complexes in aqueous solution. I. Dependence on pH. Biochim. et Biophys. Acta 278, 66—88 (1972).

BLAUER, G., HARMATZ, D.: Optical properties of bilirubin-serum albumin complexes in aqueous solution. II. Effects of electrolytes and concentration. Biochim. et Biophys. Acta 278, 89—100 (1972).

BLUM, A. D., UHLENBECK, O. C., TINOCO, I.: Circular dichroism study of nine species of transfer ribonucleic acid. Biochemistry 11, 3248—3256 (1972).

BOLARD, J., GARNIER, A.: Circular dichroism studies of myoglobin and cytochrom c derivatives. Biochim. et Biophys. Acta 263, 535—549 (1972).

BOZHKOV, V. M., KUSHNER, V. P.: Optical activity reference points of α-, β-, and unordered forms of polypeptide chains in the analysis of the ORD of globular proteins. Doklady Akad. Nauk S.S.S.R. 206, 475—478 (1972).

BREEZE, R. H., KE, B.: A circular dichroism spectrophotometer using an elasto-optic modulator. Analyt. Biochem. 50, 281—303 (1972).

BRESLOW, E., WEIS, J.: Contribution of tyrosine to CD changes accompanying neurophysin-hormone interaction. Biochemistry 11, 3474—3482 (1972).

BROWN, F. R. III, DI CORATO, A., LORENZI, G. P., BLOUT, E. R.: Synthesis and structural studies of two collagen analogues: poly(L-prolyl-L-seryl-glycyl) and poly(L-prolyl-L-alanyl-glycyl). J. Mol. Biol. 63, 85—99 (1972).

BRUGMAN, T. M., ARQUILLA, E. R.: Circular dichroism and immunologic studies of structure relationships of insulin and derivatives. Biochemistry 12, 727—732 (1973).

BUDZYNSKI, A. Z.: Circular dichroism studies on coat proteins of some strains and mutants of tobacco mosaic virus. Biochim. et Biophys. Acta 251, 292—302 (1971).

BUNTING, J. R., ATHEY, T. W., CATHOU, R. E.: Backbone folding of immunoglobulin light and heavy chains: a comparison of predicted β-bend positions. Biochim. et Biophys. Acta 285, 60—71 (1972).

BURKE, M. J., ROUGVIE, M. A.: Cross-β protein structures. I. Insulin fibrils. Biochemistry 11, 2435—2439 (1972).

BURNOTTE, J., STOLLAR, B. D., FASMAN, G. D.: Immunological and circular dichroism studies of maleylated f-1(A) histone and complexes with DNA. Arch. Biochem. Biophys. 155, 428—435 (1973).

CAMMACK, K. A., MARLBOROUGH, D. I., MILLER, D. S.: Physical properties and subunit structure of L-asparaginase isolated from Erwinia carotovora. Biochem. J. 126, 361—379 (1972).

CASEY, J. P., MARTIN, R. B.: Disulfide stereochemistry. Conformations and chiroptical properties of L-cystine derivatives. J. Am. Chem. Soc. 94, 6141—6151 (1972).

CHANDRASEKARAN, R., LAKSHIMINARAYANAN, A. V., PANDYA, U. V., RAMACHANDRAN, G. N.: Conformation of the LL and LD hairpin bends with internal hydrogen bonds in proteins and peptides. Biochim. et Biophys. Acta 303, 14—27 (1973).

CHAU, K. H., YANG, J. T.: Comparison of circular dichrometers: normal and difference CD measurements. Analyt. Biochem. 46, 616—623 (1972).

CHEN, Y.-H., YANG, J. T., MARTINEZ, H. M.: Determination of the secondary structures of proteins by circular dichroism and ORD. Biochemistry 11, 4120—4131 (1972).

CHIRGADZE, Y. N., OVSEPYAN, A. M.: Conformational transitions in globular proteins upon hydration (in Russian). Molek. Biologiya 6, 721—725 (1972).

COWBURN, D. A., BREW, K., GRATZER, W. B.: An analysis of the circular dichroism of the lysozyme-α-lactoglobulin group of proteins. Biochemistry 11, 1228—1234 (1972).

CRABBÉ, P., PARKER, A. C.: Optical rotatory dispersion and circular dichroism, in: Physical Methods of Chemistry (A. WEISSBERGER and B. W. ROSSITER, eds.), 4th ed., part IIIC pp. 183—270. New York-London-Sydney-Toronto: Wiley-Interscience 1972.

CRAWFORD, J. L., LIPSCOMB, W. N., SCHELLMAN, CH. G.: The reverse turn as a polypeptide conformation in globular proteins. Proc. Nat. Acad. Sci. US 70, 538—542 (1973).

D'ANNA, J. A., TOLLIN, G.: Studies of flavin-protein interaction in flavoproteins using fluorescence and circular dichroism. Biochemistry 11, 1073—1080 (1972).

DANIEL, E., YANG, J. T.: Analysis of the circular dichroism of the complexes of 8-anilino-1-naphthalenesulfonate with bovine serum albumin. Biochemistry 12, 508—512 (1973).

DORRINGTON, K. J., BENNICH, H., TURNER, M. W.: Conformational studies on subfragments from the Fc region of human immunoglobulin G. Biochem. Biophys. Res. Communs. 47, 512—516 (1972).

EDELMAN, G. M., CUNNINGHAM, B. A., REEKE, G. N., BECKER, J. W., WAXDAL, M. J., WANG, J. L.: The covalent and three-dimensional structure of concanavalin A. Proc. Nat. Acad. Sci. US 69, 2580—2584 (1972).

GHOSE, A.: Conformational studies on the constant region halves of heavy and light chains of human immunoglobulin G. Biochim. et Biophys. Acta 278, 337—343 (1972).
GOTTO, A. M., LUX, S. E., GOODMAN, D. S.: Circular dichroic studies of human plasma retinol-binding protein and prealbumin. Biochim. et Biophys. Acta 271, 429—435 (1972).
s'-GRAVENMADE, E. J., VAN DER DRIFT, C., VOGELS, G. D.: Conformation of allantoicase in aqueous solution. Biochim. et Biophys. Acta 251, 393—406 (1971).
GRIFFIN, J. H., ROSENBUSCH, J. P., WEBER, K. K., BLOUT, E. R.: Conformational changes in aspartate transcarbamylase. I. Studies of ligand binding and of subunit interactions by CD spectroscopy. J. Biol. Chem. 247, 6482—6490 (1972).
GROSSE, R., MALUR, J., REPKE, K. R. H.: Determination of secondary structure in isolated or membrane proteins by computer curve-fitting analysis of infrared and CD spectra. FEBS Letters 25, 313—315 (1972).

HARMISON, C. R., FROHMAN, C. E.: Conformational variation in a human plasma lipoprotein. Biochemistry 11, 4985—4993 (1972).
HELLER, W. (revised by G. CURME and W. HELLER): Optical rotation — Experimental techniques and physical optics, in: Physical Methods of Chemistry (A. WEISSBERGER and B. W. ROSSITER, eds.), 4th ed., part IIIC, pp. 51—181. New York-London-Sydney-Toronto: Wiley-Interscience 1972.
HIRATA, F., NAKAZAWA, A., NOZAKI, M., HAYAISHI, O.: Studies on metapyrocatechase. IV. Circular dichroism and ORD. J. Biol. Chem. 246, 5882—5887 (1971).
HONIG, B., KAHN, P., EBREY, T. G.: Intrinsic optical activity of retinal isomers. Implications for circular dichroism spectrum of rhodopsin. Biochemistry 12, 1637—1643 (1973).
HUSTON, J. S., BJÖRK, I., TANFORD, C.: Properties of the Fd fragment from rabbit immunoglobulin G. Biochemistry 11, 4256—4262 (1972).

JI, T. H.: Circular dichroism of a membrane protein of Neurospora crassa. Biochem. Biophys. Res. Communs. 51, 829—835 (1973).
JIRGENSONS, B.: Reconstructive denaturation of immunoglobulins by sodium dodecyl sulfate. Circular dichroism studies. Biochim. et Biophys. Acta 317, 131—138 (1973).
JONES, J. M., CREETH, J. M., KEKWICK, R. A.: Thiol reduction of human α_2-macroglobulin. The subunit structure. Biochem. J. 127, 187—197 (1972).

KALB, A. J., PECHT, I.: Visible absorption and circular dichroism of the cobalt complexes of concanavalin A. Biochim. et Biophys. Acta 303, 264—268 (1973).
KARLSSON, F. A., BJÖRK, I., BERGGÅRD, I.: Recovery of the native conformation of the variable and constant halves of an immunoglobulin light chain upon renaturation from the linear random coil state. Immunochemistry 9, 1129—1138 (1972).
KOCHWA, S., TERRY, W. D., CAPRA, J. D., YANG, N. L.: Structural studies of immunoglobulin E. I. Physicochemical studies of the IgE molecule. Ann. N. Y. Acad. Sci. 190, 49—70 (1971).
KOIDE, T., IKENAKA, T.: Studies on soybean trypsin inhibitors. 3. Amino acid sequence of the carboxy-terminal region and the complete amino acid sequence of soybean trypsin inhibitor (Kunitz). Eur. J. Biochem. 32, 417—431 (1973).

LEE, J. J., COWGER, M. L.: Circular dichroism studies of protein-bound biliverdin. Res. Commun. Chem. Pathol. Pharmacol. 5, 505—514 (1973).
LEUWENKROON-STROSBERG, E., LAASBERG, L. H., HEDLEY-WHYTE, J.: Myosin conformation and enzyme activity: Effect of chloroform, diethyl ether, and halothane on ORD and ATPase. Biochim. et Biophys. Acta 295, 178—186 (1973).
LEWIS, P. N., MOMANY, F. A., SCHERAGA, H. A.: Chain reversal in proteins. Biochim. et Biophys. Acta 303, 211—229 (1973).

LITMAN, B. J.: Effect of light scattering on the CD of biological membranes. Biochemistry 11, 3243—3247 (1972).

LITMAN, G. W., FROMMEL, D., ROSENBERG, A., GOOD, R. A.: Circular dichroic analysis of immunoglobulins in phylogenetic perspective. Biochim. et Biophys. Acta 236, 647—654 (1971).

LITMAN, G. W., LITMAN, R. S., GOOD, R. A., ROSENBERG, A.: Molecular dissection of immuno-globulin G. Conformational interrelationships of the subunits of human immunoglobulin G. Biochemistry 12, 2004—2011 (1973).

MAEDA, H., SHIRAISHI, H., ONODERA, S., ISHIDA, N.: Conformation of antibiotic protein, neo-carzinostatin, studied by plane polarized infrared spectroscopy, circular dichroism, and ORD. Intern. J. Pep. Prot. Res. 5, 19—26 (1973).

MANDEL, R., HOLZWARTH, G.: Ultraviolet CD of polyproline and oriented collagen. Bio-polymers 12, 655—674 (1973).

MATHIS, P., SAUER, K.: Circular dichroism studies on the structure and the photochemistry of protochlorophyllide and chlorophyllide holochrome. Biochim. et Biophys. Acta 267, 498 to 511 (1972).

McCUBBIN, W. D., OIKAWA, K., KAY, C. M.: Circular dichroism studies on concanavalin A. Biochem. Biophys. Res. Communs. 43, 666—674 (1971).

MELKI, G.: Étude en dichroisme circulaire de quelques hemoglobins anormales. Biochim. et Biophys. Acta 263, 226—243 (1972).

MERZ, W. E., HILGENFELDT, U., BROCKERHOFF, P., BROSSMER, R.: The time course of recombi-nation of human chorionic gonadotropin subunits studied with immunological methods, circular dichroism measurements and bioassay. Eur. J. Biochem. 35, 297—306 (1973).

MOORE, W. V., WETLAUFER, D. B.: Circular dichroism of nerve membrane fractions. J. Neuro-chem. 20, 135—149 (1973).

MORRISETT, J. D., BROOMFIELD, C. A.: Active site spin-labeled α-chymotrypsin. Guanidine hydrochloride denaturation studies using electron paramagnetic resonance and CD. J. Am. Chem. Soc. 93, 7297—7304 (1971).

MORRISETT, J. D., DAVID, J. S. K., POWNALL, H. J., GOTTO, A. M., Jr.: Interaction of an apolipoprotein (apoLP-alanine) with phosphatidylcholine. Biochemistry 12, 1290—1299 (1973).

MUKHERJEE, S., MARCHESSAULT, R. H., SARKO, A.: Far ultraviolet optical activity of saccharide derivatives. L. Xylan and cellulose acetates. Biopolymers 11, 291—301 (1972).

MYER, Y. P.: Circular dichroism studies of N-bromosuccinimide-modified horse heart cyto-chrome c preparations. Biochemistry 11, 4203—4208 (1972).

NAGANO, K.: Logical analysis of the mechanism of protein folding. I. Prediction of helices, loops, and β-structures from primary structure. J. Mol. Biol. 75, 401—420 (1973).

NAKAE, Y., IKEDA, K., AZUMA, T., HAMAGUCHI, K.: Circular dichroism of hen egg-white lyso-zyme modified with N-acetylimidazole. J. Biochem. (Tokyo) 72, 1155—1162 (1972).

NICKERSON, K. W., VAN HOLDE, K. E.: Circular dichroism study of copper(II)-ribonuclease complexes. J. Biol. Chem. 248, 2022—2030 (1973).

OLSON, J. M., PHILIPSON, K. D., SAUER, K,: Circular dichroism and absorption spectra of bacteriochlorophyll-protein and reaction center complexes from Chlorobium thiosulfato-philum. Biochim. et Biophys. Acta 292, 206—217 (1973).

PARELLO, J., PÉCHÈRE, J.-F.: Conformational studies on muscular parvalbumins. I. ORD and CD analysis. Biochimie 53, 1079—1083 (1971).

PFLUMM, M. N., WANG, J. L., EDELMAN, G. M.: Conformational changes in concanavalin A. J. Biol. Chem. 246, 4369—4370 (1971).

PHILIPSON, K. D., SAUER, K.: Comparative study of the CD spectra of reaction centers from several photosynthetic bacteria. Biochemistry 12, 535—539 (1973).

POHL, F. M.: Kooperative Konformationsänderungen von globulären Proteinen. Angew. Chem. 84, 931—944 (1972).

POLJAK, R. J.: X-ray crystallographic studies of immunoglobulins, in: Contemporary topics in molecular immunology, vol. 2 (eds. R. A. REISFELD and W. J. MANDY), pp. 1—26. New York-London: Plenum Press 1973.

PUETT, D.: Conformational studies on a glycosylated bovine pancreatic ribonuclease. J. Biol. Chem. 248, 3566—3572 (1973).

REES, D. A.: Shapely polysaccharides. Biochem. J. 126, 257—273 (1972).

RISLER, J.-L., GROUDINSKY, O.: Magnetic-CD studies of cytochrome c and cytochrome b₂. Eur. J. Biochem. 35, 201—205 (1973).

ROCKEY, J. H., MONTGOMERY, P. C., UNDERDOWN, B. J., DORRINGTON, K. J.: Circular dichroism studies on interaction of haptens with MOPC-315 and MOPC-460 mouse myeloma proteins and specific antibodies. Biochemistry 11, 3172—3181 (1972).

ROELS, H., PRÉAUX, G., LONTIE, R.: Polarimetric and chromatographic investigation of irreversible transformations of β-lactoglobulins A and B upon alkaline denaturation. Biochimie 53, 1085—1093 (1971).

ROSS, D. L., MARRACK, J. R.: The optical rotatory dispersion of antigen-antibody complexes. Immunology 23, 375—394 (1972).

ROTTEM, S., HAYFLICK, L.: Circular dichroism analysis of native and reaggregated mycoplasma membranes. Canad. J. Biochem. 51, 632—636 (1973).

SAXENA, V. P., WETLAUFER, D. B.: A new basis for interpreting the CD spectra of proteins. Proc. Nat. Acad. Sci. US 68, 969—972 (1971).

SCHNEIDER, A. B., EDELHOCH, H.: Polypeptide hormone interaction. III. Conformational changes of glucagon bound to lysolecithin. J. Biol. Chem. 247, 4992—4995 (1972).

SCHNEIDER, M. J. T., SCHNEIDER, A. S.: Water in biological membranes: adsorption isotherms and CD as a function of hydration. J. Membrane Biol. 9, 127—140 (1972).

SCHNELLBACHER, E., LUMPER, L.: CD- und ORD Spectren von Cytochrom b₅. Untersuchungen am Cytochrom b₅ aus der Mikrosomfraktion von Schweineleber. Hoppe-Seyler's Z. physiol. Chem. 352, 615—628 (1971).

SHIH, T. Y., LAKE, R. S.: Studies on the structure of metaphase and interphase chromatin of chinese hamster cells by CD and thermal denaturation. Biochemistry 11, 4811—4817 (1972).

SIMONS, E. R.: Conformational changes in some ribonuclease-inhibitor complexes. Biochim. et Biophys. Acta 251, 126—131 (1971).

SLAYTER, H. S., SHIH, T. Y., ADLER, A. J., FASMAN, G. D.: Electron microscopy and CD studies on chromatin. Biochemistry 11, 3044—3054 (1972).

STIGBRAND, T., SJÖHOLM, I.: Circular dichroism studies on the copper protein umecyanin. Biochim. et Biophys. Acta 263, 244—257 (1972).

STOREY, B. T., LEE, C. P.: Circular dichroism of cytochrome oxidase, cytochrome b₅₆₆ and cytochrome c in beef heart mitochondrial membrane fragments. Biochim. et Biophys. Acta 292, 554—565 (1973).

STRICKLAND, E. H.: Interactions contributing to the tyrosyl CD bands of ribonucleases S and A. Biochemistry 11, 3465—3474 (1972).

SUTHERLAND, J. C., SALMEEN, I., SUN, A. S. K., KLEIN, M. P.: Ferredoxin: the uses of natural and magnetic CD in a multi-chromophoric system. Biochim. et Biophys. Acta 263, 550 to 554 (1972).

TAKAGI, T., ITO, N.: Marked change of circular dichroism of L-cystine solutions with temperature. Biochim. et Biophys. Acta 257, 1—10 (1972).

TAMBURRO, A. M., JORI, G., VIDALI, G., SCATTURIN, A., SACCOMANI, G.: Studies on the structure in solutions of α-lactalbumin. Biochim. et Biophys. Acta 263, 704—713 (1972).

TANAKA, H.: The helix content of tropomyosin and the interaction between tropomyosin and F-actin under various conditions. Biochim. et Biophys. Acta 278, 556—566 (1972).

TANFORD, C.: Hydrophobic free energy, micelle formation and the association of proteins with amphiphiles. J. Mol. Biol. 67, 59—74 (1972).

TIFFANY, M. L., KRIMM, S.: Effect of temperature on the circular dichroism spectra of polypeptides in the extended state. Biopolymers 11, 2309—2316 (1972).

TIFFANY, M. L., KRIMM, S.: Extended conformations of polypeptides and proteins in urea and guanidine hydrochloride. Biopolymers 12, 575—587 (1973).

TOMIMATSU, Y., VICKERY, L. E.: Circular dichroism studies of human serum transferrin and chicken ovotransferrin and their copper complexes. Biochim. et Biophys. Acta 285, 72—83 (1972).

ULLRICH, J., WOLLMER, A.: Yeast pyruvate decarboxylase: spectral studies on the recombination of the apoenzyme with thiamine pyrophosphate and Mg. Hoppe-Seyler's Z. Physiol. Chem. 352, 1635—1644 (1971).

ULMER, D. D., HOLMQUIST, B., VALLEE, B. L.: Magnetic circular dichroism of nonheme iron proteins. Biochem. Biophys. Res. Communs. 51, 1054—1061 (1973).

VILLANUEVA, G. B., HERSKOVITS, T. T.: Exposure of tyrosyl and tryptophyl residues in trypsin and trypsinogen. Biochemistry 10, 3358—3365 (1971).

VINCENT, J.-P., CHICHEPORTICHE, R., LAZDUNSKI, M.: The conformational properties of the basic pancreatic trypsin inhibitor. Eur. J. Biochem. 23, 401—411 (1971).

WILLIAMS, R. E., LURQUIN, P. F., SELIGY, V. L.: Circular dichroism of avian erythrocyte chromatin and ethidium bromide bound to chromatin. Eur. J. Biochem. 29, 426—432 (1972).

WOLLMER, A.: Konformationsanalyse von Proteinen mit Hilfe des Circulardichroismus und der optischen Rotationsdispersion. Möglichkeiten und Grenzen, aufgezeigt an einem speziellen Hämoglobin. Habilitationsschrift, Techn. Hochschule Aachen, 1972.

WOOD, E., DALGLEISCH, D., BANNISTER, W.: Bovine erythrocyte cupro-zinc protein. II. Physicochemical properties and CD. Eur. J. Biochem. 18, 187—193 (1971).

WOODS, E. F., PONT, M. J.: Conformation changes in α-helical muscle proteins after reaction with maleic anhydride. Intern. J. Pep. Prot. Res. 4, 273—279 (1972).

Subject Index